T0118257

כתבי האקדמיה הלאומית הישראלית למדעים

PUBLICATIONS OF THE ISRAEL ACADEMY

OF SCIENCES AND HUMANITIES

SECTION OF SCIENCES

THE LICHENS OF ISRAEL

1. *Ramalina duriaei* (× 1)

2. *Ramalina reagens* (× 0.5)

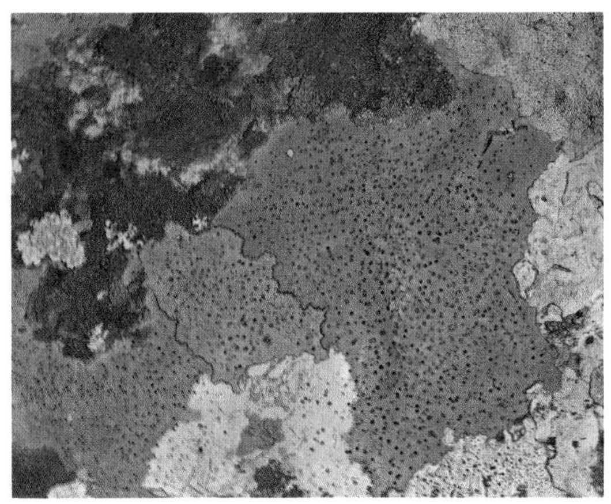

3. *Verrucaria marmorea* (× 2)

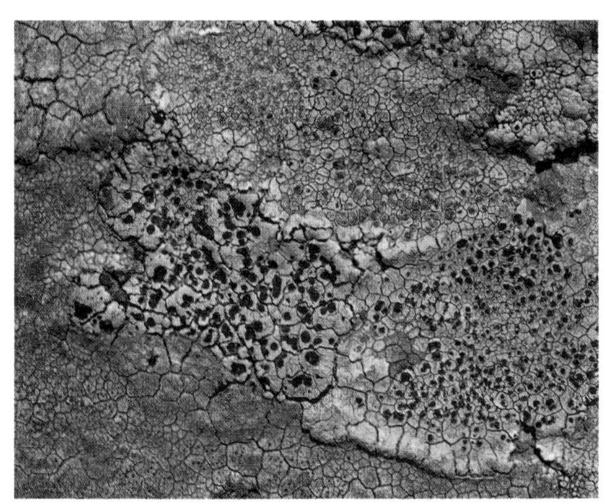

4. *Catillaria reichertiana* + *Lecanora subplanata* (× 1)

5. *Lecidea decipiens* (× 0.5)

6. *Xanthoria aureola* (× 0.8)

THE LICHENS OF ISRAEL

BY

MARGALITH GALUN

JERUSALEM 1970

THE ISRAEL ACADEMY OF SCIENCES AND HUMANITIES

Printed in Israel
At Goldberg's Press and Litho-Offset Ziv, Jerusalem
Plates by Emil Pikovsky Ltd., Jerusalem

Dedicated to

PROFESSOR ISRAEL REICHERT

PREFACE

THIS BOOK is intended to serve as an introduction to the study of the lichens of Israel and as a guide to their identification. Research in this field was initiated relatively recently (1936) by the pioneering work of Professor Israel Reichert.

For the sequential order of the families and genera, I have followed, with only minor deviations, the scheme proposed by Hale (1967).

The descriptions of species and varieties are based on my own investigations. The majority of these descriptions are illustrated by photographs, and in some cases by drawings. For the classification of the families and genera I have used the full range of diagnostic characters as they appear in the relevant literature, even though they are not always fully represented in the particular species surveyed.

The distribution recorded is based on data from the pertinent literature.

The two maps and the terms employed for recording the local and general phytogeographical distribution of the species have been taken from M. Zohary and N. Feinbrun-Dothan, *Flora Palaestina,* Jerusalem 1966→.

A special effort has been made to identify the chemical constituents of some of the lichens. These investigations were made by the usual techniques described in the works pertaining to this field (spot tests, microcrystal tests, paper chromatography and thin layer chromatography).

This study does not claim to be exhaustive. I am aware of a number of shortcomings, especially in two difficult groups — the Verrucariaceae and the subgenus *Aspicilia.* The comprehensive survey of these two groups would have been a prohibitively lengthy task.

I would like to express my sincere appreciation to all those who have helped me in various ways to complete this work, and especially to Professor I. Reichert, who introduced me to the field of lichenology and accorded me his invaluable guidance in the study presented in this volume. I have had the advantage of being able to use freely the facilities of the Farlow Reference Library and Herbarium; for this my sincere thanks are due to Dr I. M. Lamb. I am very grateful to Dr M. E. Hale for advice and criticism of the manuscript and for his aid in applying the techniques of lichen chemistry. Thanks are due to the curators of the many herbaria that sent me material on loan. I acknowledge gratefully my debt to Professor J. Poelt, Dr O. Klement, Miss H. Baumgärtner and Professor W. A. Weber for their help in clarifying some intricate material.

The photographs were taken by Mr A. Shub and the drawings were done by Mr S. Schaeffer. Their contribution to this work is considerable and I record my gratitude to them. My acknowledgements are also extended to the Israel Academy of Sciences and Humanities, which provided financial support for this research and published this volume under its auspices. I wish in particular to express my gratitude to Mr R. Eshel, the Director of Publications at the Academy, for his effective advice and guidance, and to Mrs M. Bivas, who edited the manuscript.

M. GALUN

Tel Aviv, 1970

CONTENTS

PREFACE 7

ABBREVIATIONS 10

KEY TO THE FAMILIES 11

THE FAMILIES

Collemataceae 15 • Pyrenopsidaceae 19 • Heppiaceae 22 • Pannariaceae 22 • Coccocarpiaceae 23 • Peltigeraceae 24 • Placynthiaceae 25 • Lecideaceae 26 • Cladoniaceae 36 • Diploschistaceae 37 • Pertusariaceae 42 • Acarosporaceae 44 • Lecanoraceae 47 • Parmeliaceae 63 • Usneaceae 66 • Physciaceae 69 • Teloschistaceae 83 • Verrucariaceae 96 • Dermatocarpaceae 98 • Cypheliaceae 102 • Arthoniaceae 102 • Arthopyreniaceae 103 • Opegraphaceae 104 • Dirinaceae 105 •

GLOSSARY 107

BIBLIOGRAPHY 110

INDEX 112

PLATES *after p.* 116

MAPS *at end*

ABBREVIATIONS

B	$Ba(OH)_2$ (a saturated aqueous solution of barium hydroxide)
C	$Ca(OCl)_2$ (a saturated aqueous solution of calcium hypochlorite)
C.	central
E.	east, eastern
GAAn	glycerine–95% alcohol–aniline (2 : 2 : 1)
GAoT	glycerine–alcohol–*o*-toluidine (2 : 2 : 1)
GAW	glycerine–alcohol–water (1 : 1 : 1)
GE	glycerine–acetic acid (1 : 1)
I	I–KI (iodine in potassium iodide solution)
K	KOH (a 10% aqueous solution of potassium hydroxide)
KC	a combination of solutions K and C (K is applied first, followed by C)
N.	north, northern
Pd	$C_6H_4(NH_2)_2$ (a solution of 5% *p*-phenylenediamine in alcohol)
S.	south, southern
W.	west, western

KEY TO THE FAMILIES

1. Lichens with open disc- or cup-shaped fruiting bodies (apothecia) 2
- Lichens with immersed, roundish or flask-shaped fruiting bodies (perithecia) 28
2. Asci disintegrating at maturity and spores liberated to form a mazaedium. **Cypheliaceae**
- Not as above 3
3. Phycobiont belonging to the Cyanophyceae (blue-green algae) 4
- Phycobiont belonging to the Chlorophyceae (green algae) 10
4. Thallus homoeomerous, gelatinous when moist 5
- Thallus heteromerous, not or only slightly gelatinous when moist 7
5. Algae unicellular, solitary or aggregated in colonies (Chroococcales). **Pyrenopsidaceae**
- Algae filamentous 6
6. Algae in simple chains of globose cells (*Nostoc*) (Fig. 1 J). **Collemataceae**
- Algae in branched filaments (*Stigonema* (Fig. 1 I) or *Hypomorpha*). **Coccocarpiaceae**
7. Thallus foliose, usually large. **Peltigeraceae**
- Thallus crustose, small-squamulose or small-foliose 8
8. Spores 2- to multicellular. **Placynthiaceae**
- Spores simple 9
9. Asci with 8 spores. Phycobiont *Nostoc* (Fig. 1 J). **Pannariaceae**
- Asci with 4 to many spores. Phycobiont *Scytonema* (Fig. 1 H). **Heppiaceae**
10. Algae unicellular, globose, mainly *Trebouxia* (Fig. 1 A) 14
— Algae multicellular filamentous — *Trentepohlia* (Fig. 1 E) 11
11. Fruiting bodies closed but for an apical pore. **Arthopyreniaceae**
— Fruiting bodies open, discoid, elongate or irregular 12
12. Apothecia immarginate. **Arthoniaceae**
— Apothecia marginate 13
13. Apothecia with exciple and thalline margin. **Dirinaceae**
— Apothecia with exciple only. **Opegraphaceae**
14. Lichens composed of a horizontal, crustose or squamulose thallus and vertical, hollow stalks (podetia). **Cladoniaceae**
— Not as above 15
15. Asci with 1–32 spores 16
— Asci with more than 32 spores. **Acarosporaceae**
16. Apothecia enclosed in thalline warts. Spores thick-walled, more than 30 μ long. **Pertusariaceae**
— Apothecia not in warts. Spores thin-walled, usually less than 30 μ long 17
17. Apothecia with thalline margin, with or without exciple (lecanorine) 18
— Apothecia with exciple and without thalline margin (lecideine or biatorine) 24
18. Spores pigmented 19
— Spores colourless 20
19. Spores brown, transversely septate, thin- or thick-walled. **Physciaceae**
— Spores green to brown, muriform. **Diploschistaceae**
20. Spores placodiomorph; if spores simple or septate then epithecium and thallus K+ purple. **Teloschistaceae**
— Spores simple or septate, thin-walled. Epithecium and thallus not K+ purple 21

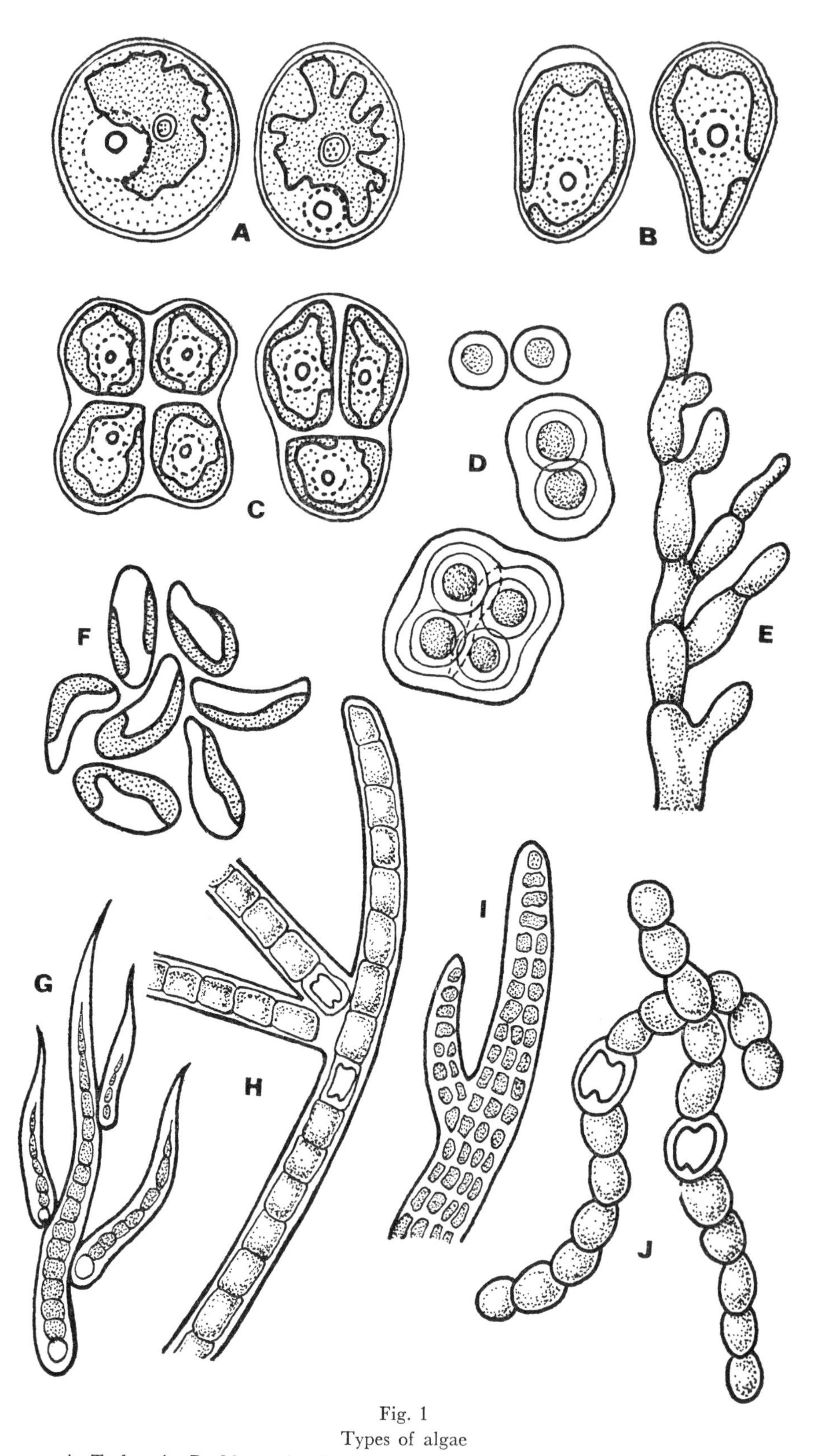

Fig. 1
Types of algae
A. *Trebouxia*; B. *Myrmecia*; C. *Pleurococcus*; D. *Gloeocapsa*; E. *Trentepohlia*; F. *Coccomyxa*; G. *Rivularia*; H. *Scytonema*; I. *Stigonema*; J. *Nostoc*
(A, B, C, E, F, H, J after V. Ahmadjian, 1967; D, G, I after C. F. E. Ericksen, 1957)

21. Thallus fruticose. **Usneaceae**
— Thallus not as above 22
22. Thallus foliose or subfruticose. **Parmeliaceae**
— Thallus crustose, uniform, lobate-effigurate or squamulose 23
23. Thallus yellow or orange, K–. **Candelariaceae**
— Thallus variously coloured, K+ or K–. **Lecanoraceae**
24. Spores colourless 25
— Spores pigmented 26
25. Spores placodiomorph. **Teloschistaceae**
— Spores simple or septate, thin-walled. **Lecideaceae**
26. Disc urceolate. **Diploschistaceae**
— Disc plane, convex or concave 27
27. Spores septate or muriform. Thallus yellow to green. *Rhizocarpon* (**Lecideaceae**)
— Spores 2–4 septate. Thallus not yellow or green. **Physciaceae**
28. Thallus crustose, superficial or growing within the substrate, ecorticate. **Verrucariaceae**
— Thallus foliose, squamulose or subcrustose, well developed, distinctly corticate above or on both sides. **Dermatocarpaceae**

COLLEMATACEAE

Thallus gelatinous, unstratified, granular, subfruticose or foliose. Phycobiont *Nostoc*. Apothecia lecanorine. Asci with 4–8 spores. Spores colourless or yellowish to pale brown, simple or variously septate.

1. Spores simple. **2. Physma**
– Spores septate to muriform. **1. Collema**

1. Collema G. H. Web.

Thallus foliose, olive-green to blackish, homoeomerous, fragile when dry, and when moist rapidly absorbing large amounts of water and swelling.

Apothecia with a red to reddish brown disc, an unstratified thalline margin and a cellular (euthy-, subpara- or euparaplectenchymatous) exciple (Fig. 2 A–C). Spores usually colourless, sometimes faintly coloured, septate to muriform.

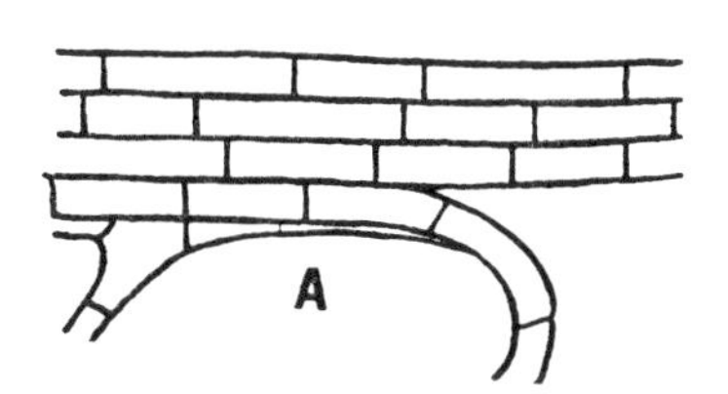

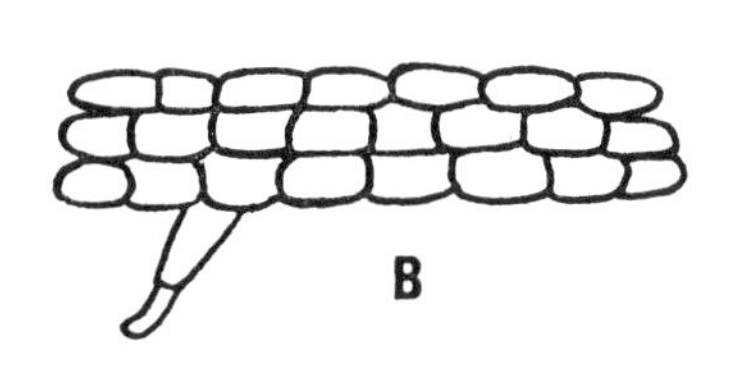

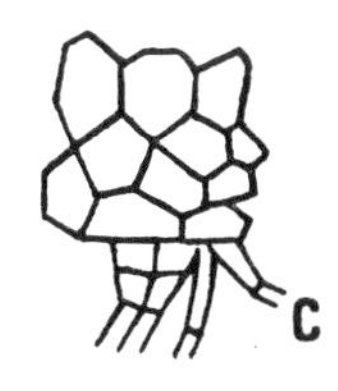

Fig. 2
The three types of proper margin in *Collema*
A. euthyplectenchymatous; B. subparaplectenchymatous; C. euparaplectenchymatous
(after Degelius, 1954)

1. Thallus isidiate 2
– Thallus without isidia 5
2. Spores submuriform, with both ends or one end acute. **6. C. tunaeforme**
– Spores with transverse septae only (rarely mixed with some submuriform spores) 3
3. Spores long and narrow, acicular, multiseptate. **3. C. nigrescens**
– Spores usually 3-septate, some 1–2-septate, few submuriform 4
4. Spores with both ends rounded or one end acute. Isidia globular-granular. **5. C. tenax** var. **vulgare**
– Spores with obtuse or rounded ends. Isidia squamiform. **1. C. crispum**
5. Spores submuriform with both ends acute. Exciple euparaplectenchymatous. **2. C. cristatum**
– Spores usually 3-septate, some 1–2-septate, mixed with some submuriform spores. Exciple not as above 6
6. Exciple distinctly euthyplectenchymatous. **5. C. tenax** var. **vulgare**
– Exciple eupara- to subparaplectenchymatous. **4. C. polycarpon**

1. Collema crispum (Huds.) G. H. Web., in Wiggers, Primit. Fl. Holsat. 89 (1780). *Lichen crispus* Huds., Fl. Anglica ed. 1, 447 (1762).

Thallus foliose, dark green to black (paler when wet), varying in size; lobes either broad, roundish-lobate, rather uniformly thick, growing in rosettes, or reduced to form a 'corona' around the apothecia; margins ascending, undulate to cup-shaped. Isidia squamiform, sparse or numerous, mainly restricted to the thalline margin of mature apothecia.

Apothecia numerous, single or crowded, *c.* 1 mm across; adnate. Disc plane, reddish brown, slightly glossy. Thalline margin concolourous with the thallus but remaining dark when wet, somewhat prominent, at first entire, later isidiate. Exciple thin, pale yellowish, not well separated from the hymenium, subparaplectenchymatous beneath the hymenium and rather euthyplectenchymatous laterally. Hymenium colourless. Epithecium yellowish brown, gelatinous. Spores 8, colourless, transversely 3-septate, oval or ovoid with obtuse or rounded ends, sometimes constricted at the septae, 28–32 × 12–13.5 μ.

Reactions: Thallus in section and exciple I–; hymenium I+ blue.

Habitat: On sandy soil and on calcareous rocks. Upper Galilee, N. Negev. (Specimens identified by Degelius were from Mt. Carmel, Judean Mountains and Judean Desert.)

Distribution: Medio-European, Mediterranean, Irano-Turanian, W. and E. Saharo-Arabian; also found in N. America.

2. Collema cristatum (L.) G. H. Web., in Wiggers Primit. Fl. Holsat. 89 (1780). *Lichen cristatus* L., Spec. Plant. 1143 (1753).

Thallus foliose, dark olive-green, slightly glistening, semi-transparent when wet, several cm across, roundish or irregular in shape, adnate with simple rhizines; lobes contiguous and imbricate or discrete, short and broad, 150–300 μ thick, repeatedly branched; margins entire or crenulate. Isidia absent.

Apothecia numerous, marginal or laminal, sessile, up to 2 mm across. Disc plane or concave, blackish brown, slightly glossy. Thalline margin rather thick, entire, prominent, persistent. Exciple euparaplectenchymatous. Hymenium colourless. Epithecium yellowish brown. Subhymenium yellowish. Spores 8, colourless or yellowish brown, submuriform, with 3–4 transverse and 1 longitudinal septae; both ends acute; 18–35 × 8–12 μ.

Reactions: Hymenium and subhymenium I+ blue; exciple I–.

Habitat: On calcareous rocks, on soil and among mosses. Upper Galilee, Mt. Carmel. (Specimens identified by Degelius were from the Upper Jordan Valley, Mt. Carmel and Judean Desert.)

Distribution: Common in the N. Hemisphere.

3. Collema nigrescens (Huds.) DC., in Lamarck and De Candolle, Fl. Franc. ed. 3, 2: 384 (1805). *Lichen nigrescens* Huds., Fl. Anglica ed. 1, 450 (1762). [Plate I: 1]

Thallus foliose-membranaceous, broadly lobate, appearing in more or less roundish colonies several cm across, dark greenish black to almost black (old parts yellowish brown), lower side somewhat paler; lobes roundish, closely adnate to the substrate except for the somewhat ascending, entire, usually undulate margins; upper surface strongly ridged and pustulated; ridges radiating or nearly so, often densely isidiated. Isidia small, granular.

Apothecia numerous, mostly concentrated on the ridges, sessile to short and broadly stipitate, up to 1 mm across. Disc plane, at first pale brown then dark brown. Thalline margin thick, entire, at first prominent then partly rolled back. Hymenium colourless with brownish, gelatinous epithecium. Exciple pale yellowish, sub- to euparaplectenchymatous. Spores 8, colourless, long and narrow-acicular to bacillar, 50–65 × 2.5–3.5 μ, 6–14 celled.

Reactions: Hymenium I+ blue; exciple I–.

Habitat: On bark of olive trees. Upper Galilee.

Distribution: N. America, Atlantic, Mediterranean; N. W. Boreal penetrating Medio-European territories.

4. Collema polycarpon Hoffm., Deutschl. Flora 102 (1796).

Thallus foliose, slightly irregular in shape, up to 6 cm across, adnate, olive-green to blackish, dull, deeply lobate and lobulate; lobes radiating or nearly so, contiguous, 1–5 mm broad, more or less flattened, coarsely plicate; edges swollen, raised; lower side somewhat paler with rounded hapters of whitish rhizines.

Apothecia sparse or numerous, sessile, 1–3 mm across. Disc plane, reddish brown, pruinose. Thalline margin entire, thin or thick, slightly prominent at first then almost disappearing. Exciple eupara- to subparaplectenchymatous, colourless below and faintly yellowish above. Hymenium 70–120 μ thick, colourless except for the brownish, gelatinous epithecium. Paraphyses simple or furcate; 2–4 uppermost cells thickened and pigmented. Asci usually with 8 spores. Spores colourless, 4-celled or 2–3-celled, with one or both ends acute, 17–20.5 × 3.5–6.8 μ.

Reactions: Thallus in section I+ wine-red; hymenium I+ blue; exciple I–.

Habitat: On calcareous rocks, soil and basalt. Upper and Lower Galilee, Judean Mountains. (Specimens identified by Degelius were from Upper Galilee and Mt. Carmel, both var. *corcyrense*.)

Distribution: Common in the N. Hemisphere.

5. Collema tenax (Sw.) Ach., var. **vulgare** (Schaer.) Degel., Symb. Bot. Upsal. 13(2): 163 (1954). *C. pulposum* var. *vulgare* Schaer., Enum. Crit. Lich. Europ. 259 (1850).

Thallus foliose, irregularly spreading and up to 5 cm across, or in the form of roundish cushions 2–3 cm across; lobes imbricate or separate, 0.3–1 cm long, auriculate, narrow below, broadened and raised above, attached to the substrate by groups of thin, white rhizines; blackish or dark green above, pale green to almost transparent below, somewhat bluish when moist, dull, epruinose; margins usually undulate-lobulate with 2–5 (–10) roundish lobules. Isidia present or absent.

Apothecia rare to very many, adnate, up to 3 mm across. Disc reddish brown, plane. Thalline margin rather thin, entire. Exciple euthyplectenchymatous. Hymenium 80–

150 μ thick, colourless with brownish gelatinous epithecium. Asci usually with 8 spores. Spores colourless, usually 3-septate, some 1-septate and rarely submuriform; both ends rounded or one end acute; 15–22 × 6–8 μ.

Reactions: Thallus in section I+ wine-red; hymenium I+ blue.

Distribution: Medio-European, Mediterranean extending into adjacent Irano-Anatolian territories.

F. **vulgare** Degel., Symb. Bot. Upsal. 13(2): 163 (1954) [Plate I: 2]. Thallus in form of roundish cushions, no isidia.

Habitat: On plaster of ruins and on loess. C. and W. Negev. (Specimens identified by Degelius were from the Judean Mountains.)

F. **papulosum** (Schaer.) Degel., Symb. Bot. Upsal. 13(2): 163 (1954). *C. pulposum α vulgare papulosum* Schaer., Enum. Crit. Lich. Europ. 259 (1850). Thallus roundish or irregular, with many globular isidia.

Habitat: On soil in fissures of calcareous rocks and on mossy soil. Mt. Carmel.

6. Collema tunaeforme (Ach.) Ach., Lichgr. Univ. 649 (1810). *Parmelia tunaeformis* Ach., Meth. Lich. 227 (1803).

Thallus foliose, roundish or irregular in shape, up to 5 cm across, dark olive-green to almost black (paler when moist), adnate to substrate but easily detachable; lobes roundish, undulate, deeply and broadly lobate; margins ascending, discrete or imbricate. Isidia numerous, globular.

Apothecia numerous to crowded, sessile, 0.5–2 mm across. Disc plane, reddish brown. Thalline margin thin, entire or isidiate. Hymenium colourless, 80–120 μ thick. Subhymenium yellowish. Epithecium gelatinous, yellowish. Exciple distinctly euparaplectenchymatous, colourless. Spores 8, colourless, submuriform with 3 transverse septae and one longitudinal septum, slightly constricted at the septae; one or both ends acute; 20–29 × 7.5–10 μ.

Reactions: Thallus in section I+ wine-red; hymenium I+ blue; spores I+ reddish blue; exciple I–.

Habitat: On bare calcareous rocks and on mossy surfaces. Upper Galilee, Mt. Carmel, Judean Mountains.

Distribution: In temperate and arctic zones of the N. Hemisphere.

2. PHYSMA Mass.

Thallus foliose, fragile when dry, swelling when moistened, homoeomerous. Phycobiont *Nostoc,* in chains. Apothecia lecanorine. Thalline margin unstratified. Exciple cellular, colourless. Spores colourless, simple, thick-walled.

1. Physma omphalarioides (Anzi) Arn., Flora 50: 119 (1851). *Collema omphalarioides* Anzi, Comment. Soc. Crittog. Ital. 1 (3): 131 (1862). [Plate I: 4]

Thallus foliose, several cm across, thin, dark green, paler when moist, central part adnate; periphery lobate; margin entire, flat and adnate or loose and ascending;

upper surface markedly wrinkled or plicate, densely covered with granular isidia.

Apothecia sparse or numerous, 1–1.5 mm across, constricted at the base. Disc plane, dark reddish brown. Thalline margin at first entire, later incised and isidiate (Fig. 3). Exciple euparaplectenchymatous. Hymenium colourless, 120–160 μ thick. Epithecium gelatinous, yellowish. Spores 8, colourless, simple, oval, 15–20 × 9–12 μ.

Habitat: On the bark of the lower shaded parts of *Quercus calliprinos* trunks. Upper Galilee.

Distribution: Mediterranean; also in N. Norway.

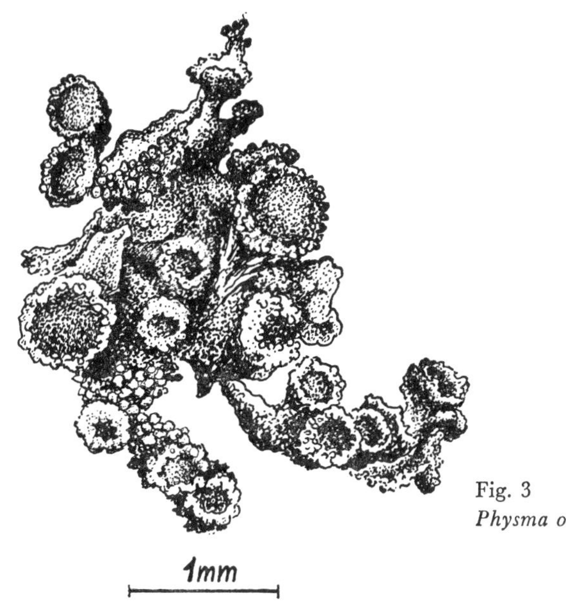

Fig. 3
Physma omphalarioides

PYRENOPSIDACEAE*

Thallus crustose or minutely squamulose, foliose or minutely fruticulose, blackish, dark green or dark brown, turgescent, homoeomerous. Phycobiont blue-green algae belonging to the Chroococcales. Ascocarps nearly closed to discoid, of the stroma type or simple, usually lecanorine. Asci with 8 to many spores. Spores usually single-celled, colourless.

1. Apothecia containing algae in the hymenium. **1. Gonohymenia mesopotamica**
- Apothecia without algal cells in the hymenium 2
2. Thallus crustose-granulose. **3. Psorotichia numidella**
- Thallus foliose or fruticulose 3
3. Thallus fruticulose of minute, ramose, more or less cylindrical branches with apothecia. **2. Peccania** sp.
- Thallus monophyllous, in a contiguous crust, sterile. **4. Thyrea** sp.

* According to Zahlbruckner (1926).

1. Gonohymenia Stein.

Thallus squamulose, homoeomerous. Algae xanthocapsoid. Apothecia of the stroma type or simple, containing algal cells in the hymenium. Asci polysporous. Spores colourless, simple, thin-walled.

1. Gonohymenia mesopotamica Stein., Ann. Naturhist. Mus. Wien 34:21 (1921). [Plate II : 1]

Thallus externally resembling *Collema,* black when dry, turgescent and olive-brown when wet, squamulose, affixed to the substrate by hyphal strands; squamules erect or nearly so, 2.5–3 mm high and 1.5–2.5 mm broad, rotund or flat, irregularly plicate; tips usually broadened to form an almost flattened surface. Algae densely and more or less perpendicularly aggregated along the periphery of the squamules, gradually dispersing toward the centre but absent in the central medulla; algal cells embedded in a gelatinous sheet 2–3 μ thick, which is yellow near the outermost cells and colourless near the inner cells; algal cells usually in pairs or in groups of 4, yellow with a faint pinkish tinge; each cell penetrated by a hyphal haustorium. Medullary hyphae 0.8–1 μ thick, hyaline, anastomose, loose and approximately parallel in arrangement.

Apothecia almost confluent with the flattened tips of the squamules and concolourous with them, consequently distinguishable only by microscopic examination. Disc up to 900 μ in diameter, with irregular, delicate cracks and without a distinct thalline margin. Apothecia of the stroma type, separating sterile regions, 70–80 μ broad, composed of algal cells and medullary hyphae (Pl. II : 2). Paraphyses thin, flaccid, branching. Asci polysporous. Spores simple, oval to roundish, 4–5.5 μ.

Habitat: On basalt. Upper Jordan Valley.

Distribution : Reported only from Mesopotamia.

2. Peccania (Mass.) Forss.

Thallus fruticulose, composed of tiny cylindrical to flattened ramose branches; homoeomerous. Phycobiont *Gloeocapsa* (Fig. 1 D). Apothecia terminal, discoid, with a thick thalline margin. Hypothecium colourless. Asci with 8 or many spores. Spores simple, colourless, thin-walled.

1. Peccania sp. (Fig. 4)

Thallus a more or less contiguous crust of blackish squamules arranged in clusters, 0.5–2 mm broad and *c.* 1 mm high; squamules 150–350 μ thick, ramose, erect, papillate, subcylindrical to cylindrical, yellowish in section, connected with the substrate by central strands of colourless hyphae. Algal cells usually single, concentrated mainly at the periphery in a zone 40–50 μ thick, centre consisting of colourless hyphae parallel to the surface.

Apothecia *c.* 0.5 mm across, lecanorine. Disc blackish, granular, naked. Thalline margin prominent, incurved, thick. Hymenium colourless with a thick, gelatinous,

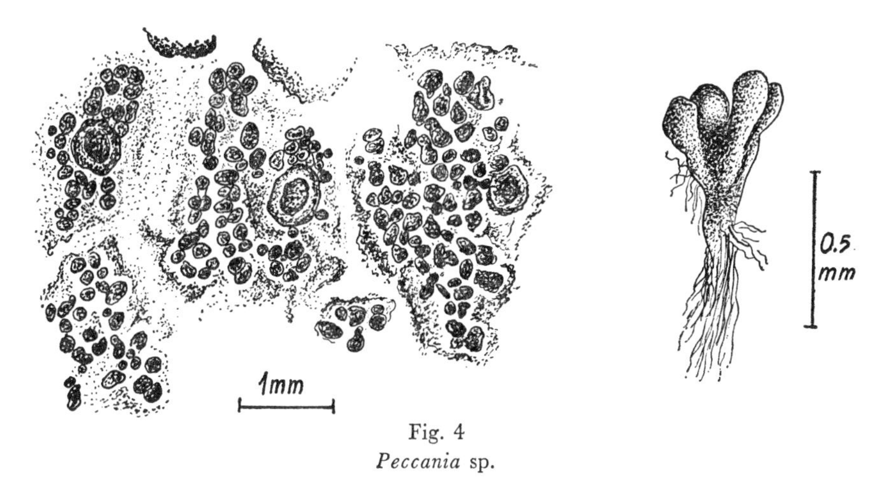

Fig. 4
Peccania sp.

dark brown epithecium. Hypothecium yellowish. Spores 8, simple, subglobose, 7.5–9 μ.
Reactions: Thallus in section K–; hymenium and hypothecium I+ blue.
Habitat: On thin layers of loess that accumulate on Nubian sandstone. C. Negev.

3. PSOROTICHIA Mass.

Thallus crustose-granulose or minutely squamulose, homoeomerous. Phycobiont *Gloeocapsa* (Fig. 1 D). Apothecia innate-urceolate, with a thick, connivent thalline margin. Asci usually with 8 spores. Spores simple, colourless, thin-walled.

1. Psorotichia numidella Forss., Nova Acta Reg. Soc. Scient. Upsal. ser. 3, 13 : 76 (1885). [Plate I : 3]

Thallus homoeomerous, composed of more or less dispersed granules, 0.2–1 mm, black (dark brown when wet), partially immersed at first and later emergent; exterior a gelatinous yellowish brown layer 6–9 μ thick and inspersed with minute yellow granules. Algal cells in large groups concentrated mainly near the periphery.

Apothecia urceolate, small. Disc at first almost invisible, later plane, up to 350 μ across, brown. Thalline margin thick, tumid. Hymenium colourless, 120–140 μ thick. Epithecium yellowish brown, granular. Hypothecium colourless, 40–50 μ thick. Paraphyses septate, unbranched, uniform. Spores 8, colourless, simple, subglobose or ovoid, thin-walled, 7.5–10.5 × 6–7.5 μ.

Reactions: Hymenium I+ blue; algal cells K+ brown.
Habitat: On dolomite and crystalline limestone. W. and C. Negev, Arava Valley.
Distribution: Known from N. Africa and Lusitania.

4. THYREA Mass.

Thallus foliose, small, monophyllous, umbilicate, usually in a contiguous crust, homoeomerous. Phycobiont *Gloeocapsa* (Fig. 1 D). Apothecia lecanorine, with a thick thalline margin. Spores 8–24, simple, colourless.

1. Thyrea sp.

Thallus rosette-shaped, consisting of substipitate lobes 1–2 mm broad and up to 2 mm high; lobes blackish, ascending, irregularly plicate, crenulate, surface furfuraceous-isidiose. Algal cells solitary or in groups of 2–4, densely concentrated at the periphery, dispersed towards the centre. Central hyphae more or less parallel to the surface.

Sterile.

Habitat: On thin layers of fine sandy soil in crevices of bare granitic rocks. Arava Valley.

HEPPIACEAE

Thallus subfoliose to squamulose, paraplectenchymatous. Phycobiont *Scytonema* (Fig. 1 H). Apothecia immersed. Asci with 4 to many spores. Spores colourless, simple.

Heppia Naeg.

Thallus squamulose, entirely or partly paraplectenchymatous, composed of large cells. Phycobiont *Scytonema* cells in roundish colonies. Apothecia immersed (at least at first). Hypothecium colourless. Asci with 4 or many spores. Spores colourless, simple.

1. Heppia psammophila Nyl., Flora 61 : 339 (1878).

Thallus squamulose; squamules 1–3 mm across and *c.* 0.6 mm thick, contiguous; upper surface olive or dark brown, opaque, at first roundish and plane, soon becoming concave with crenulate margins and finally deformed with cracked surface; lower side sand-coloured, attached by central rhizoidal hyphae 5–15 μ thick; upper layer distinctly paraplectenchymatous, consisting of perpendicular cell rows and covered by a thick, amorphous stratum. Algae arranged in dense groups. Medulla of colourless, more or less longitudinal hyphae, delimited below by groups of roundish cells.

Apothecia rare, solitary or 2–3 per squamule, up to 1.5 mm across, at first immersed, later adnate. Disc plane, brown. Hymenium 150–180 μ thick, colourless. Epithecium yellowish, gelatinous. Hypothecium cellular, faintly yellow, subtended by algal cells. Paraphyses slender, coherent, simple. Spores numerous, colourless, spherical, 5–7 μ in diameter.

Reactions: Hymenium I+ pink; epithecium K+ purple.

Habitat: On thin layers of fine sandy soil accumulated in crevices of granite rocks in Elat.

Distribution: Reported from Liguria (Italy).

PANNARIACEAE

Thallus squamulose or foliose, stratified. Hypothallus usually present. Phycobiont *Nostoc* (Fig. 1 J). Apothecia lecanorine or biatorine. Asci with 8 colourless simple spores.

PANNARIA Del.

Thallus small-foliose or squamulose, in shades of grey or brown. Upper cortex paraplectenchymatous. Phycobiont *Nostoc,* in chains or colonies mainly in the upper part of the white medulla. Lower cortex lacking. Medulla usually delimited below by thick, pigmented, longitudinal hyphae. Apothecia lecanorine. Hypothecium colourless or pale. Spores 8, simple, colourless.

1. **Pannaria mediterranea** C. Tav., Portug. Acta Biol. (B) 8 : 5 (1965).

Thallus partly remaining in the form of small lobulate squamules up to 3 mm across, but mostly becoming at first marginally and later almost entirely granulose-sorediate so that the squamulose nature of the thallus is disguised; squamules greyish brown. Soredia lead-grey. Upper cortex paraplectenchymatous, faintly yellowish, 25–35 μ thick, covered with a yellowish amorphous stratum and subtended by *Nostoc* colonies which also penetrate the loosely aggregated colourless medulla. Lower side composed of brownish hyphae more or less parallel to the surface. Hypothallus bluish black.

Sterile.

Reactions : Thallus K–, C–, Pd–.

Habitat : On bark of *Quercus calliprinos.* Upper Galilee.

Distribution : Reported from France, Portugal and Morocco.

COCCOCARPIACEAE

Thallus foliose or fruticose, attached to the substrate by rhizines. Phycobiont filamentous, blue-green. Apothecia brown or black, lecideine. Spores 8, simple, colourless, oval or spherical. Paraphyses thick, septate, branched.

SPILONEMA Born.

Thallus fruticose, minute, tree-like or irregularly branched, rhizinate. Phycobiont *Stigonema* or *Hypomorpha.* Apothecia sessile, brown or black, lecideine. Spores 8, simple, small, oval, colourless.

1. **Spilonema revertens** Nyl., Flora 48 : 601 (1865).

Thallus composed of blackish cushions, 3–6 mm wide and up to 4 mm high, consisting of irregularly branched, rigid, densely aggregated filaments with greenish blue rhizines. Phycobiont *Stigonema* (Fig. 1 I).

Sterile.

Habitat : On basalt. Upper Jordan Valley.

Distribution : Temperate regions of the N. Hemisphere.

Thallus foliose, stratified. Phycobiont blue-green (*Nostoc*) (Fig. 1 J) or green (*Coccomyxa*) (Fig. 1 F) or both. Apothecia immarginate.

NEPHROMA Ach.

Thallus foliose, greenish yellow or brown; upper and lower cortices paraplectenchymatous. Phycobiont *Nostoc* or *Coccomyxa*. Apothecia on the lower surface of the lobal tips, immersed. Spores 8, 3-septate, light brown.

1. Nephroma laevigatum Ach., Synops. Lich. 242 (1814). [Fig. 5]

Thallus foliose, large, horizontally spreading; upper side brown, glabrous, here and there slightly wrinkled, with few marginal isidia; lower side pale brown, wrinkled, slightly pubescent, 140–220 μ thick. Upper cortex colourless. Phycobiont *Nostoc,* in a more or less uniform layer. Medulla partly white and partly yellow. Lower cortex yellow with densely packed outgrowths of yellow hyphae.

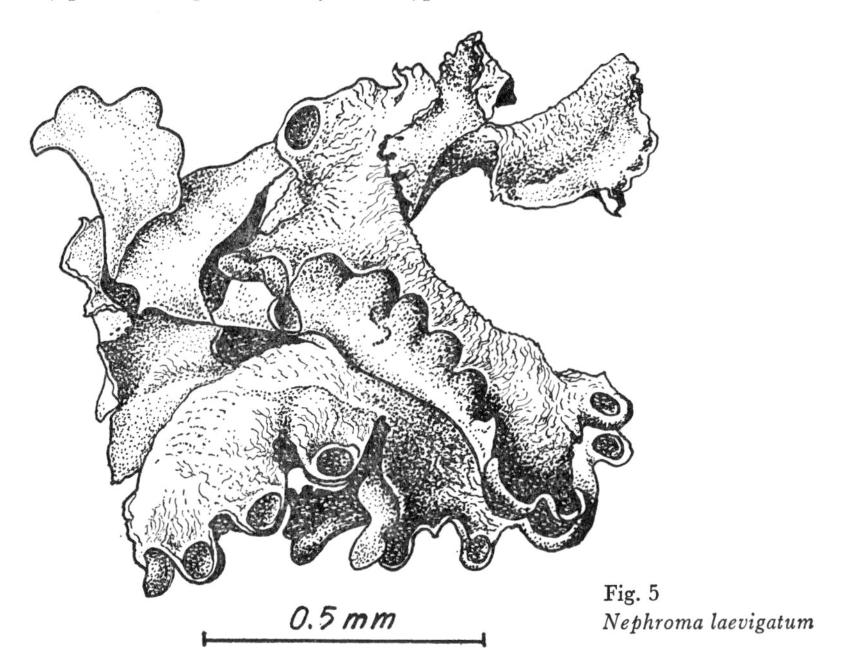

Fig. 5
Nephroma laevigatum

Apothecia numerous, up to 4 mm in diameter, on the lower surface at the lobal tips, finally directed upwards by the deflexing of the fertile lobes. Disc brown (paler than the upper side), plane, roundish, immarginate. Hymenium colourless. Epithecium yellowish brown, gelatinous. Spores pale brown, 3-septate, ellipsoid to ovoid, with thickened walls, 17–20 × 5–6 μ.

Reactions: Medulla K+ purple; contains nephromin (Pl. XXVII : 3).

Habitat: On the lower part of trunks that are shaded most of the time by dense surrounding vegetation. Upper Galilee.
Distribution: In oceanic areas (Wetmore, 1960).

PLACYNTHIACEAE

Thallus crustose to minutely coralloid-squamulose. Algae filamentous, blue-green, belonging to the Rivulariaceae or Scytonemataceae. Spores 2- to multicellular.

PLACYNTHIUM (Ach.) S. F. Gray*

Thallus grey, olive or black, granular-crustose or cracked-areolate or minutely coralloid-squamulose, usually with a bluish green hypothallus; hyphal arrangement pseudo-parenchymatous-reticulate. Phycobiont belonging to Rivulariaceae or Scytonemataceae. Apothecia with exciple or thalline margin. Hymenium green or brown above. Spores 8, 2–4-celled, colourless, ellipsoid or fusiform.

1. Placynthium nigrum (Huds.) S. F. Gray, A Nat. Arrang. Brit. Plants 1: 395 (1821). *Lichen niger* Huds., Fl. Anglica ed. 2, 2: 524 (1778).

Thallus appearing in blackish patches composed of papillose or flat, erect or adpressed squamules, 0.3–1.5 mm thick and 0.3–3 mm high; squamules scattered at the periphery and densely crowded in the centre, forming an areolated crust with the areoles 1–3 mm across. Hypothallus bluish green; hyphae 4–6 μ thick, more or less radially arranged, emerging at the circumference as a marginal zone *c.* 1 mm broad. Phycobiont *Dichotrix orsiniana* (Geitler, 1934), solitary or in filaments, uniformly distributed on a reticulum of fungal cells. Lower cortex composed of more or less longitudinal hyphae parallel to the surface.

Apothecia 0.5–2 mm in diameter. Disc black or brownish black, at first concave, later convex. Exciple black, shiny, finally disappearing, composed of a cellular (pseudo-parenchymatous) violet-brown tissue, 50–70 μ thick below. Hypothecium yellowish brown, 70–100 μ thick. Paraphyses branched and septate, with the apical cells thickened. Spores 8, colourless, ellipsoid, straight or slightly curved, 2–4-celled, 9–16 × 4–6 μ.

Habitat: On mortar, dolomite and soft calcareous rocks. Upper Galilee, Coast of Carmel, Judean Mountains.
Distribution: Common in temperate regions of the N. Hemisphere.

* Placed by Henssen (1963) in the Peltigeraceae.

LECIDEACEAE

Thallus crustose, uniform, lobate-effigurate, squamulose or obsolete, stratified, attached to the substrate by medullary or hypothalline hyphae. Upper cortex well differentiated, scarcely apparent, or lacking. Phycobiont belonging to the Chlorophyceae. Apothecia discoid, superficial or partly immersed. Exciple colourless to carbonaceous. Thalline margin absent. Paraphyses simple or branched and intricate. Spores usually 8, colourless or green to brown (in *Rhizocarpon*), simple, septate or muriform.

1. Spores colourless, simple or transversely septate — 2
– Spores at first colourless, later olive-green to brown, septate to muriform. — **5. Rhizocarpon**
2. Spores simple — 3
– Spores septate — 4
3. Exciple externally and internally pigmented. Epithecium gelatinous. — **3. Lecidea**
– Exciple internally colourless. Epithecium granular. — **4. Problastenia**
4. Thallus squamulose. — **6. Toninia**
– Thallus crustose to obsolete — 5
5. Spores 1-septate. — **2. Catillaria**
– Spores multiseptate. — **1. Bacidia**

1. Bacidia De Not.

Thallus crustose, effuse, usually poorly developed. Apothecia superficial. Exciple pale or dark. Thalline margin absent. Hypothecium colourless, pale or brown but never black. Spores usually 8, colourless, multiseptate, elongate, acicular.

1. Bacidia albescens (Kremp.) Zw., Flora 45 : 495 (1862). *Scoliciosporum molle* f. *albescens* Kremp., Denkschr. Kgl. Bayer. Bot. Ges. 4 (2) : 207 (1861).

Thallus of fine, greenish, rather inconspicuous granules.

Apothecia fairly numerous, sessile, irregularly dispersed or several crowded together, up to 1.2 mm across. Disc at first pale, later darker reddish brown, naked or slightly pruinose, at first plane, sometimes moderately convex later. Exciple paler than the disc, pruinose or naked, at first thick and prominent, later excluded. Exciple colourless and clear in section, composed of parallel radiating hyphae. Hymenium clear and colourless with a yellowish granular epithecium. Hypothecium colourless, compactly hyphose, variable in thickness, with inspersed groups of algae. Paraphyses simple and slender. Spores 8, colourless, multiseptate with only transverse septae, elongate-acicular, straight, 40–56 × 3 μ.

Reactions: Thallus and disc K–; exciple K+ brown; hymenium I+ blue.

Habitat: On bark of *Quercus calliprinos* and *Pinus halepensis*. Upper Galilee, Judean Mountains.

Distribution: Frequent in the N. Hemisphere.

2. CATILLARIA (Ach.) Th. Fr.

Thallus crustose, uniform, thick to almost lacking, attached to the substrate by medullary hyphae. Cortex prosoplectenchymatous. Phycobiont belonging to the Chlorococcales. Apothecia partly immersed or superficial. Exciple dark. Thalline margin absent. Hypothecium colourless, yellowish to brown or black. Spores 8, colourless, 2-celled or often remaining single-celled, with thin cell walls and septa.

1.	Corticolous.	**1. C. chalybeia** f. **ilices**
–	Saxicolous	2
2.	Thallus poorly developed, K–.	**2. C. piciloides**
–	Thallus well developed, K+ yellow.	**3. C. reichertiana**

1. Catillaria chalybeia (Borr.) Mass. f. **ilicis** (Mass.) Vain., Acta Soc. Fauna et Flora Fenn. 57 (2): 430 (1954). *Lecidea ilicis* Mass., Mem. Lich. 124 (1853).

Thallus very thin, minutely areolate, pale ash-grey, very poorly developed.

Apothecia 0.3–0.5 mm across, adnate. Disc black, naked, at first plane, later moderately convex. Exciple externally black, thin, at first prominent, later partly excluded, composed of conglutinate brownish hyphae. Hymenium colourless except for a thin, fuliginous, gelatinous epithecium. Hypothecium brownish. Paraphyses coherent; apical cells clavate with brownish cell walls. Spores 8, colourless, 1-septate, ellipsoid, 8–10 × 3–3.5 μ.

Reactions: Thallus K–, C–, Pd–; hymenium I+ blue turning wine-red.

Habitat: On twigs of olive trees. Mt. Carmel, Judean Mountains.

2. Catillaria piciloides Zahlbr., Verh. Zool.-Bot. Ges. Wien 67 : 16 (1917).

Thallus ash-grey, very thin and almost invisible.

Apothecia sparse, up to 1 mm across, rotund, adnate to partially immersed, leaving shallow reddish pits in the substrate when removed. Disc dark brown or black, plane, naked. Exciple black, slightly prominent, 30–50 μ thick. Hymenium dark brown, 50–70 μ thick. Spores 8, colourless, 1-septate, oblong-ellipsoid, 16–17 × 5–8 μ.

Reactions: Thallus K–, C–; asci I+ blue turning greenish.

Habitat: On oolithic limestone. Mt. Carmel.

Distribution: Medio-European.

3. Catillaria reichertiana Galun, The Lichenologist 3 : 423 (1967). [Plate II : 3]

Thallus crustose, more or less determinate, 1–6 cm across, pale greenish grey, up to 3 (3.5) mm thick, unequally rimose-areolate; areoles plane or convex, sometimes with low protuberances and becoming almost verruculose; lower side white, easily detachable from substrate, sometimes delimited by a blackish hypothallus. Cortex very thin, prosoplectenchymatous, with many inspersed crystals. Algal layer continuous. Medulla partly opaquely greyish due to numerous crystals, partly brownish with reddish brown particles.

Apothecia (Fig. 6) sparse to numerous, dispersed to crowded, immarginate, rounded or irregular, at first immersed, eventually becoming sessile, 1–1.5 mm across. Disc

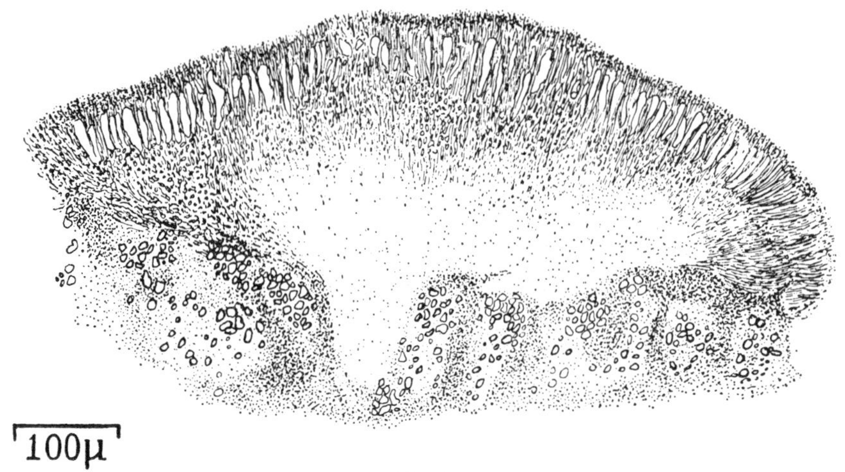

Fig. 6
Catillaria reichertiana; transverse section of ascocarp
(from Galun, 1967a)

plane, black or brownish black. Hymenium colourless, 25–40 μ thick. Epithecium greenish, gelatinous, covered by a fine yellow granular layer. Hypothecium colourless or sometimes the upper part colourless and the lower part yellowish. Exciple thin, indistinct or absent. Paraphyses uniform; apices with an external gelatinous, greenish sheath. Spores 8, simple or becoming 2-celled with very thin septa, colourless, ellipsoid to subglobose, 4.5–11 × 2–5 μ.

Reactions: Thallus K+ yellow turning yellow-brown; KC+ deep yellow, P–, C–; contains usnic acid (Pl. XXV: 4). With I the spores turn green and the apical part of the asci and the subhymenial region blue, while all other parts remain unchanged.

Habitat: On calcareous rocks in exposed positions. Always in association with *Lecanora atra*. Upper Galilee, C. Negev.

3. Lecidea Ach.

Thallus crustose, uniform or lobate-effigurate or squamulose, stratified; attached to the substrate by medullary or hypothalline hyphae. Phycobiont usually *Trebouxia* (Fig. 1 A). Apothecia discoid, superficial or partly immersed, lecideine. Disc dark brown or black. Paraphyses simple. Spores 8, colourless, simple ellipsoid or ovoid.

Subgen. EULECIDEA Stzbgr. Thallus variously crustose.

1. Corticolous — 2
– Saxicolous — 3
2. Thallus C+ red. Hypothallus black. — **4. L. olivacea**
– Thallus C–. Hypothallus absent. — **2. L. euphorea**
3. Hypothecium colourless. — **3. L. ocellulata**
– Hypothecium brown — 4
4. Thallus greyish, C–. — **1. L. algeriensis**
– Thallus yellowish green, C+ orange. — **5. L. subincongrua** var. **elaechromoides**

1. Lecidea algeriensis (Flag.) Zahlbr., Cat. Lich. Univ. 3 : 511 (1925). *L. maculosa* Flag., Rev. Myc. 14 : 75 (1892).

Thallus crustose, indeterminate, thin, rimose-areolate; surface rough, greyish.

Apothecia abundant, scattered or crowded and often confluent, located between the areoles and level with them, up to 1 mm across. Disc black, dull, epruinose, angular, persistently plane. Hymenium 70–100 μ thick, colourless. Epithecium gelatinous, fuliginous. Hypothecium dark brown, somewhat paler in its lower part, 250–375 μ thick. Exciple black, 40–75 μ laterally, expanding to 50–150 μ above. Paraphyses septate, simple, conglutinate. Spores 8, colourless, ovoid, 4–10.5 × 3.5–5 μ.

Reactions: Thallus K–, C–, KC+ red, Pd–; hymenium I+ bluish green turning wine-red.

Habitat: On basalt. Upper Galilee.

Distribution: Reported only from Algeria.

2. Lecidea euphorea (Floerk.) Nyl., Flora 70: 187 (1881). *L. sabuletorum* var. *euphorea* Floerk., Berl. Mag. 2 : 311 (1808). [Plate III : 1]

Thallus rugose-verruculose, 100–140 μ thick; verrucules 0.1–0.3 mm across, whitish or yellowish grey, occasionally pale olive-green. Hypothallus absent.

Apothecia numerous, up to 1 mm across, adnate. Disc black, epruinose, at first plane, later convex to tuberculose and immarginate. Exciple black, glossy, in section, bluish externally and violet towards the hymenium. Hymenium 70–90 μ thick, colourless. Epithecium gelatinous, olive-green. Hypothecium yellow above, brown below. Paraphyses coherent, uniform or slightly enlarged at the apex. Spores 8, colourless, simple, ellipsoid, 10–14 × 6–7 μ.

Reactions: Thallus K+ yellowish, C–, KC+ yellow-orange; hymenium I+ blue.

Habitat: Common on oak and olive trees. Upper Galilee, Mt. Carmel, Samaria, Judean Mountains.

Distribution: Common in the N. Hemisphere.

3. Lecidea ocellulata Th. Fr., Lichgr. Scand. 1 : 484 (1874).

Thallus determinate, up to 6 cm across, 0.5 mm thick, brownish green, rimose-areolate, fissures deep and almost uniform; areoles 0.3–1 mm across, smooth, naked and slightly shiny or occasionally covered thinly with a pruina, plane or moderately concave; margin whitish, slightly raised.

Apothecia abundant, up to 0.8 mm across, immersed between the areoles, single and roundish or several crowded so as to loose their originally circular shape. Disc black, pruinose, plane or concave. Exciple black, blackish brown in section, epruinose, persistent, prominent, 20–30 μ thick. Hymenium colourless, 140–200 μ thick. Epithecium obscurely brown. Hypothecium 140–230 μ thick, composed of colourless hyphae perpendicular to the surface. Asci numerous. Spores 8, ellipsoid, 10.5–18 × 7.5–10.2 μ.

Reactions: Thallus K–, C–, KC–, Pd–; hymenium I–.

Habitat: On basalt. Lower Galilee.

Distribution: Mediterranean and Medio-European.

4. Lecidea olivacea (Hoff.) Mass., Ric. Auton. Lich. Crost. 71 (1852). *Verrucaria olivacea* Hoff., Deutschl. Fl. 2 : 192 (1796). [Plate III : 2]

Thallus in small patches, delimited by a black hypothallus, thin, verruculose, olive-green, rarely greyish.

Apothecia numerous, 0.5–1 mm across, adnate. Disc black, epruinose, at first plane and marginate, later convex and exciple excluded. Exciple macroscopically black, in section violet near the hymenium; peripheral zone bluish. Hymenium colourless, 65–70 μ thick. Epithecium gelatinous, bluish green. Hypothecium brown, paler above. Paraphyses coherent, slightly enlarged at the apex. Spores 8, simple, colourless, ovoid to ellipsoid, 9–14 × 6–7 μ.

Reactions: Thallus K+ yellow, C+ red, KC+ red; hymenium I+ blue turning wine-red.

Habitat: On olive trees. Upper and Lower Galilee, Judean Mountains.

Distribution: Common in the N. Hemisphere.

5. Lecidea subincongrua Nyl. var. **elaeochromoides** (Nyl.) Poelt, Ber. Bayer. Bot. Ges. 34 : 86 (1961). *L. parasema* var. *elaeochromoides* Nyl., Bull. Soc. Linn. Normand. 6 : 310 (1872).

Thallus indeterminate, thin, rimose-areolate; areoles warty-granular, yellowish green.

Apothecia scattered or crowded, 0.2–0.6 mm across, immersed in the areoles, sometimes emerging. Disc black, naked, plane. Exciple persistent, externally black, slightly glossy, internally yellowish brown. Hymenium 70–90 μ thick, colourless. Epithecium thick, gelatinous, bluish green. Hypothecium dark brown. Spores 8, ellipsoid, simple, colourless, 9–14 × 6–8 μ.

Reactions: Thallus K+ yellow, C+ orange.

Habitat: On basalt. Lower Galilee.

Distribution: Mediterranean.

Subgen. PSORA Th. Fr. Thallus lobate or squamulose.

1.	Thallus radiate-lobate. Medulla yellow.	**8. L. opaca**
–	Thallus squamulose. Medulla white	2
2.	Thallus brown. Medulla K+ yellow→red, Pd+ yellow.	**6. L. albilabra**
–	Thallus pink to reddish brown. Medulla K+ red, Pd–.	**7. L. decipiens**

6. Lecidea albilabra Duf. in Nyl., Act. Soc. Linn. Bordeaux 12 : 367 (1856). [Plate IV : 1]

Thallus squamulose; squamules roundish, slightly sinuate when mature, congested, up to 5 mm across, adnate, with free edges; upper side brown and white-rimmed; lower side whitish. Cortex gelatinous, yellowish, composed of somewhat disconnected perpendicular rows of roundish-ellipsoid cells (±fastigiate). Algal layer dense, continuous, more or less uniform. Medulla macroscopically white, greyish-nubilated in section.

Apothecia numerous, 1 or many on each squamule, sessile, at first plane, later convex. Disc dark brown or black, sometimes with a pruinose rim along the exciple. Hymenium 65–90 μ thick, colourless. Epithecium brown, gelatinous, 12–15 μ thick. Subhymenium distinct, faintly yellowish, 50–55 μ thick. Hypothecium 250–300 μ in depth, nubilated by dark greyish crystals. Exciple composed of colourless, parallel hyphae embedded in a thick, yellowish gelatinous stratum. Spores 8, simple, colourless, ellipsoid, 10–13 × 3–5 μ.

Reactions: Medulla K+ yellow turning red, C–, Pd+ yellow; contains norstictic acid (Pl. XXVI: 3).

Habitat: On soil in rock depressions. Judean Mountains.

Distribution: Mediterranean and Medio-European.

7. Lecidea decipiens (Ehrh.) Ach., Method. Lich. 80 (1803). *Lichen decipiens* Ehrh., Hedw., Stirp. Crypt. 2: 7 (1789). [Plate IV: 2, 3]

Thallus squamulose; squamules 2–5 mm across, 200–350 μ thick, discrete or contiguous, at first plane, smooth and roundish, later irregular with a cracked surface, firmly attached to the substrate, except for the slightly raised white margin, by hyphae 3–7.5 μ thick; upper side pink or reddish brown, epruinose or thinly or heavily pruinose; lower side whitish. Cortex yellowish, covered by a somewhat transparent amorphous layer 10–30 μ thick. Algal layer dense and continuous, interrupted only beneath the apothecia. Medulla white, but nubilated by greyish crystals.

Apothecia usually numerous, marginal, 1 or several on each mature squamule, 0.5–2 mm across. Disk black, at first plane and surrounded by a white exciple, later convex or shapeless and exciple excluded. Epithecium brown, covered by a somewhat transparent amorphous stratum. Hymenium 80–125 μ thick, pale brown. Hypothecium colourless or brownish, lower part with many crystals. Spores 8, simple, ellipsoid, 10–16 × 5–6 μ.

Reactions: Medulla K+ red, Pd–, C–; hymenium I+ blue.

Habitat: On loess, on earth in limestone crevices, or among mosses and various plant roots. Upper Galilee, Mt. Carmel, Philistean Plain, Judean Mountains, Judean Desert, N., C. and W. Negev.

Distribution: In temperate regions, steppes and semideserts of the N. Hemisphere.

8. Lecidea opaca Duf. in Fries, Lichgr. Europ. Reform. 289 (1831).

Thallus orbicular, up to several cm across, dark olive-green, distinctly radiate-plicate, firmly attached to the substrate; lobes contiguous, 3–5 mm long and 0.5–1 mm broad; centre areolate. Cortex 20–25 μ thick, paraplectenchymatous, colourless, only the uppermost cells brownish, covered in part by a thin yellowish brown amorphous layer. Algal layer more or less uniform and continuous. Medulla yellow.

Apothecia rather numerous, up to 1 mm across, sessile. Disc black or blackish brown, plane, naked. Exciple black, persistent, prominent, forming three-quarters of a circle, 100–130 μ deep and 80–100 μ thick, cellular; externally dark brown, internally colourless or somewhat fuliginous. Hypothecium 150–170 μ thick, almost reaching the substrate, colourless except for a faintly brownish centre of densely aggregated brownish cells. Hymenium colourless, 80–90 μ thick. Epithecium brown. Paraphyses coherent;

apical cells with a brownish cap. Spores 8, simple, ellipsoid, colourless, 6.5–8.5 × 3–5 μ.
Reactions: Medulla K+ brownish red, C+ red; contains erythrin (Pl. XXIV: 3).
Habitat: On calcareous rocks. Upper Galilee.
Distribution: Mediterranean.

4. PROTOBLASTENIA Stein.

Thallus crustose, uniform, partly endolithic to almost invisible; ecorticate. Apothecia discoid, surperficial or partly immersed. Exciple internally colourless. Epithecium granular. Spores 8, simple, colourless.

1. Apothecia reddish brown, K+ purple. Hypothecium colourless to pale yellow. **1. P. calva**
- Apothecia dark brown to black, K–. Hypothecium brown. **2. P. immersa**

1. Protoblastenia calva (Dicks.) Zahlbr., Cat. Lich. Univ. 7: 1 (1930). *Lichen calvus* Dicks., Fasc. Plant. Crypt. Brit. 2: 18 (1789). [Plate III: 3]

Thallus very thin, whitish-farinose, almost invisible.

Apothecia numerous, scattered or in groups, up to 1.5 mm across, round, broadly attached to the substrate. Disc very soon becoming convex to almost globose, reddish brown, dull, epruinose. Exciple very thin and soon excluded to the level of the hypothecium, composed of parallel-radiate colourless hyphae with brownish end cells. Hypothecium colourless or faintly yellowish. Hymenium colourless. Epithecium yellowish brown, granular. Paraphyses conglutinate, branched. Spores 8, colourless, simple, ovoid, 10–18 × 5.8 μ.

Reactions: Thallus K–, C–; apothecia K+ purple; hymenium I+ blue.

Habitat: On calcareous rocks. Upper Galilee.

Distribution: Medio-European and Mediterranean.

2. Protoblastenia immersa (Web.) Stein., Verh. Zool.-Bot. Ges. Wien 65: 203 (1915). *Lichen immersum* Web., Spicil. Fl. Goetting. 188 (1778).

Thallus a continuous very thin whitish crust, partly immersed and rather indistinct.

Apothecia numerous, *c.* 1 mm across, sessile or immersed in depressions or pits. Disc dark brown to blackish, naked, soon becoming immarginate, plane or convex. Hymenium 100–150 μ thick, with a yellowish tinge. Epithecium granular, brown, 30–35 μ thick. Hypothecium pale brown. Spores 8, colourless, ellipsoid or ovate, simple or 2-celled, sometimes both simple and 2-celled spores in the same ascus, 10–17 × 6–8 μ.

Reactions: Thallus and apothecia K–; hymenium I+ reddish brown; hypothecium I+ blue.

Habitat: On calcareous rocks. Mt. Carmel.

Distribution: Common in temperate regions of the old world.

5. RHIZOCARPON Ram.

Thallus crustose, areolate with a black hypothallus, corticate above by a pseudocortex (Degelius, 1954). Phycobiont belonging to the *Cystoccoccus* type (Runemark, 1956). Apothecia situated on the hypothallus, lecideine. Disc always black. Exciple internally colourless with a black cortex. Paraphyses branched and intricate. Asci usually with 8 spores. Spores septate or muriform, first colourless and later greenish brown.

1. Rhizocarpon tinei (Tornab.) Run. ssp. **tinei** Run., Opera Bot. 2(1): 122 (1956). [Plate III : 4]

Thallus indeterminate, up to 6 cm across, bright greenish yellow, areolate; areoles 0.3–0.7 mm across, irregular in shape, plane, smooth, dull, forming, together with the apothecia, a continuous crust. Hypothallus broad, black, visible between the areoles.

Apothecia numerous, between and level with the areoles, solitary or in groups, 0.3–0.8 μ across, polygonal. Disc black, naked. Exciple thin, black, at first moderately raised, later indistinct. Hymenium colourless. Epithecium reddish brown, covered by a colourless amorphous stratum. Hypothecium blackish below, greenish brown above. Paraphyses conglutinate, branched, with brownish apical cell walls. Spores 8, oblong-ellipsoid, at first colourless, later greenish brown to blackish, pluriseptate to muriform, deformed when mature, 28–43 × 10–15 μ.

Reactions: Thallus K–, C–, Pd+ yellow; contains barbatic acid and psomoric acid (Pl. XXIV : 4); medulla and hymenium I+ blue.

Habitat: On quarzolithic and melecoid limestones and on basalt. Upper Galilee.

Distribution: Mediterranean, extending into Medio-European territories.

6. TONINIA Th. Fr.

Thallus of thick turgid or small inflated squamules, corticated above. Phycobiont belonging to the Chlorophyceae. Apothecia superficial, black, lecideine. Exciple of radiately arranged thickened hyphae. Spores 8, colourless, transversely 1- or more-septate, ellipsoid or fusiform.

1. Spores 1-septate with pointed ends — 2
– Spores usually 3-septate with rounded ends — 3
2. Squamules subglobose, greyish white, small. — **3. T. coeruleonigricans**
– Squamules plane to convex, brown with a white rim, rather large. — **1. T. albomarginata**
3. Epithecium violet. Hypothecium brown. — **2. T. aromatica**
– Epithecium reddish brown. Hypothecium colourless. — **4. T. verrucosa**

1. Toninia albomarginata B. de Lesd., Bull. Soc. Bot. France 82: 315 (1935). [Plate V : 2]

Thallus squamulose; squamules up to 5 mm across, dispersed or crowded, the entire lower surface more or less attached to the substrate; young squamules greenish brown when wet with distinct whitish edges, at maturity turning brown, the white edges more or less disappearing by growing downwards, the squamules becoming turgid and deformed and the surface cracked. Cortex 80–90 μ thick, reticular-cellular, composed of thin-walled somewhat perpendicularly orientated hyphae, covered by an amorphous stratum 150–170 μ thick invaded by thick brown parasitic (?) hyphae and many dispersed dead cells. Algal layer continuous, 50–60 μ thick. Medulla opaque with thick medullary hyphae, singly and in groups, penetrating deep into the substrate.

Apothecia fairly numerous, marginal, solitary or in groups up to 3 mm across, often appearing to be immarginate from the beginning. Disc black, densely pruinose, plane or moderately convex. Exciple entire laterally and below the disc, brown, consisting of a layer, 50–70 μ thick, of parallel, conglutinate hyphae with capitate apical cells, covered by an amorphous layer 10–20 μ thick. Upper part of hypothecium brown and densely intricate, colourless and gradually less dense towards the base. Hymenium 45–85 μ thick with a sordid brown epithecium. Paraphyses simple, rarely branched, apical cells pigmentated and clavate. Spores 8, 1-septate with pointed ends, 17–22 × 3–3.5 μ.

Reactions: Medulla K–, C–, Pd+ dark blue (mainly in the upper zone).

Habitat: On loess in shady pits among rocks. C. Negev.

Distribution: Known only from Liguria (Italy) and Morocco.

2. Toninia aromatica (Turn.) Mass., Framment. Lichen. 24 (1855). *Lecidea aromaticus* Turn., Transact. Linn. Soc. London 9: 140 (1808). [Plate V: 1]

Thallus squamulose; squamules irregular, verrucose-glebose, greyish white, pruinose, *c.* 1 mm wide, dispersed or in groups, attached to the substrate or fused with it, rarely lobate at the periphery. Cortex 25–50 μ thick, transparent with conglutinate hyphae entwined in various directions and covered by a thick granular stratum. Algal layer interrupted by strands of perpendicularly orientated cortical hyphae. Medulla thick, white, lower part opaque.

Apothecia numerous and crowded, usually larger than the squamules. Disc black, naked or slightly pruinose. Exciple externally black, disappearing, consisting of very dark brown radiate hyphae intergrading with the hypothecium. Hymenium 50–70 μ thick. Epithecium violet. Subhymenium brown, 40–50 μ thick. Spores 8, colourless, with rounded ends, usually 3-septate, sometimes 1- or 4–5-septate with cells of regular or irregular size, 17–21 × 3–3.5 μ.

Reactions: Thallus and medulla K–, C–, KC–, Pd–.

Habitat: On sandstone and on loess. Coast of Carmel, C. Negev.

Distribution: In temperate regions of the N. Hemisphere.

3. Toninia coeruleonigricans (Lightf.) Th. Fr., Lichgr. Scand. 1: 336 (1874). *Lichen coeruleonigricans* Lightf., Fl. Scotica 2: 805 (1777). [Plate V: 3]

Thallus squamulose; squamules small and thick, roundish or kidney-shaped, turgid or turgid-plicate, greyish green, appressed, scattered or contiguous, with a dense white, uniform or (in dry environment) reticulate pruina. Cortex reticulate-cellular, transparent, 15–25 μ thick, covered by an amorphous layer *c.* 15 μ thick with dispersed dead cells. Medulla in contact with a thick brownish hypothallus connected with the substrate by strands of brownish hyphae 5–7 μ thick which penetrate deep into the substrate.

Apothecia numerous, situated among the squamules, up to 2 mm across. Disc plane, black, naked or slightly pruinose. Exciple externally black, pruinose or naked, becoming excluded with growth, internally dark brown along the subhymenium and slightly paler below the disc, radiate in structure. Hymenium, including the brownish epithecium, 60–70 μ thick; upper zone of hypothecium of dark brown, densely aggregated hyphae gradually becoming less dense and colourless below. Spores 8, 1-septate, colourless, fusiform, straight or slightly curved, with many vacuoles, 15–19 × 2–3 μ.

Habitat: On earth in rock crevices. Upper Galilee, Mt. Carmel, Judean Mountains, C. Negev.

Distribution: Pluriregional.

4. **Toninia verrucosa** (Mass.) Flag., Cat. Lich. Algerie 62 (1896). *Thalloidima verrucosum* Mass., Mem. Lichgr. 122 (1853). [Fig. 7]

Thallus squamulose, squamules scattered to crowded, densely pruinose, somewhat regularly and deeply cracked, arranged in rosette-like tubercules. Cortex 25–40 μ thick; hyphae conglutinate, with a thick amorphous layer above. Algal cells grouped. Medulla entirely nubilated by granules and substrate particles.

Apothecia numerous, on and between the squamules. Disc black, pruinose, plane and marginate at first, later convex with margin excluded. Exciple externally black and dark brown in section laterally and below. Epithecium and upper zone of the hymenium reddish brown. Subhymenium pale brown. Hypothecium colourless. Paraphyses simple; apical cells enlarged (6–7 × 3–5 μ) and pigmented red-brown. Spores 8, colourless, 4-celled, straight with rounded ends, 17–24 × 3–4 μ.

Habitat: On mortar and cement. Upper Galilee.

Distribution: Medio-European and Mediterranean.

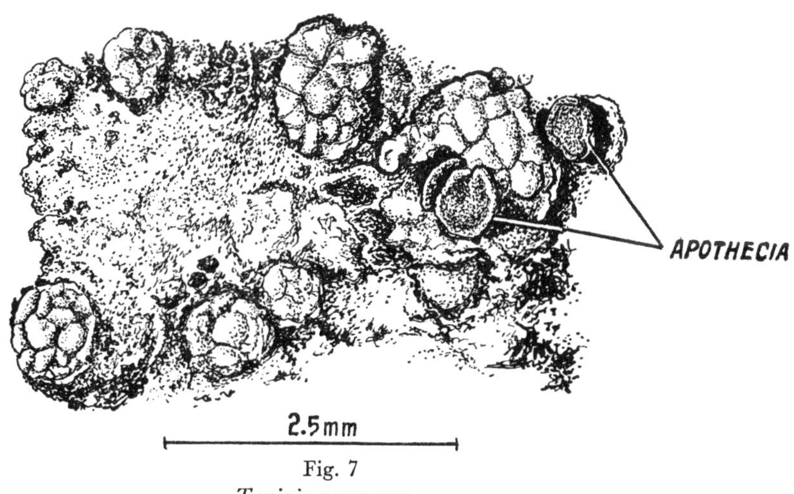

Fig. 7
Toninia verrucosa

CLADONIACEAE

Here represented only by the genus *Cladonia.*

CLADONIA Hill.

Differentiated into a horizontal primary thallus and a secondary vertical thallus, the podetium. Primary thallus crustose or squamulose, consisting of an upper cortex, algal layer and medulla; lower cortex absent. Phycobiont *Trebouxia* (Fig. 1A). Podetia arising from the margins of the primary thallus or from its upper surface, erect, hollow, simple or branched and shrubby, with pointed or cup-shaped ends. Apothecia located on the tips of the podetia, biatorine. Spores 8, simple, colourless.

1.	Medulla Pd+ red (fumarprotocetraric acid). Podetia simple	2
–	Medulla Pd–. Podetia dichotomously branched.	3. **C. rangiformis**
2.	Squamules large (1–3 cm long). Podetia small (1–3 mm tall).	1. **C. convoluta**
–	Squamules smaller (0.5–1 cm long). Podetia larger (*c.* 1 cm tall).	2. **C. pocillum**

1. Cladonia convoluta (Lam.) Cout., in H. Magn. Flora over Skandin. busk-och bladlavar 63 (1929). *Lichen convolutus* Lam., Encycl. Bot. 3 : 500 (1789). [Plate VI : 1]

Primary thallus squamulose; squamules leafy, ascending, revolute, 1–3 cm long and 0.3–0.8 cm wide, compacted into a crowded, continuous crust; margins irregularly incised to crenate-lobate; upper surface olive-green, green or brownish, more or less shiny; lower surface pale yellow, usually visible from above because of the revolute squamules.

Podetia rare, emerging from the upper surface of the squamules and concolourous with them, simple, cylindrical, or with a cup-shaped apex, 1–3 (–5) mm tall.

Apothecia 0.5–1.5 mm across, terminal and single or several on the cup margin. Disc convex, brown. Spores 8, simple, colourless, ellipsoid, 8–12 × 2–3 μ.

Reactions: Thallus K–, KC+ yellow; medulla Pd+ yellow→brick-red; contains usnic acid (Pl. XXV : 4) and fumarprotocetraric acid (Pl. XXV : 3).

Habitat: On mossy soil and dry plant remains. Upper Galilee, Mt. Carmel.

Distribution: Mediterranean and Medio-European; also found in the British Isles.

2. Cladonia pocillum (Ach.) Rich., in H. Magn., Flora over Skandin. busk-och bladlavar 61 (1929). *Baeomyces pocillum* Ach., Method. Lich. 336 (1803). [Fig. 8]

Primary thallus a continuous crust of squamules; squamules rather thick, foliose, 0.5–1 cm long and up to 4 mm wide; margins partly ascending, irregularly incised, crenate-lobate; upper surface olive-green or brownish green; lower surface white.

Podetia emerging from the upper side of the squamules, grey or greyish green, simple, erect, hollow with deep cup-shaped ends, up to 1 cm tall, overgrown with pale green, granular, circular flakes.

Apothecia rare, on the margins of the cups. Disc brown, convex. Spores 8, simple, colourless, ellipsoid, 9–14 × 3.5–4 μ.

Reactions: Thallus K–, C–; margins of medulla and squamules Pd+ red; contains fumarprotocetraric acid (Pl. XXV: 3).

Habitat: On calcareous soil in rock fissures and on mossy soil, often together with *C. convoluta*. Upper Galilee, Mt. Carmel, Coast of Carmel, Shefela, Judean Mountains.

Distribution: Cosmopolitan.

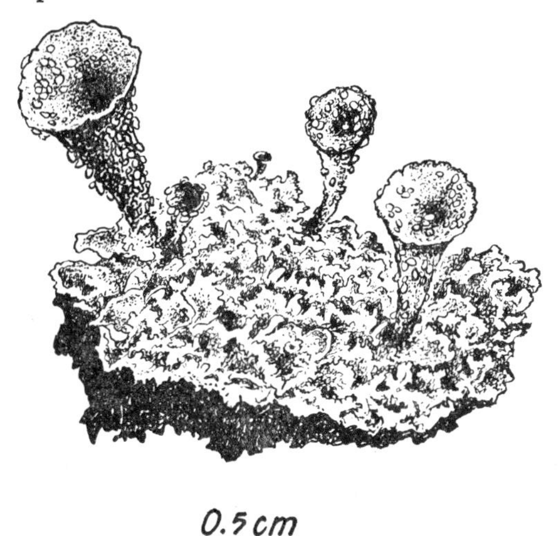

Fig. 8
Cladonia pocillum

3. Cladonia rangiformis Hoffm., Deutschl. Fl. 2: 114 (1796). [Fig. 9]

Primary thallus squamulose; squamules densely aggregated, 1–3 mm long and about as wide; margins irregularly crenate or incised-crenate, ascending and revolute; upper surface brownish or olive-green; lower surface white.

Podetia in congested tufts, up to 1 cm tall, cylindrical, repeatedly dichotomously branched, whitish with dark brown tips, densely covered with ellipsoid or roundish green spots.

Sterile.

Reactions: Thallus and podetia K+ yellow, C–, Pd–; contains atranorin (Pl. XXV: 1).

Habitat: On mossy soil. Rare. Upper Galilee.

Distribution: In Mediterranean and Medio-European lowlands; also reported from N. America.

DIPLOSCHISTACEAE

Thallus uniformly crustose, stratified, attached to the substrate by hyphothalline or medullary hyphae. Cortex poorly developed. Phycobiont belonging to the Chlorophyceae. Apothecia innate or adnate. Disc urceolate or plane. Exciple well developed, colourless or dark. Thalline margin present or lacking. Paraphyses thin and slender, usually not branched. Spores multiseptate and colourless or muriform and green to brown.

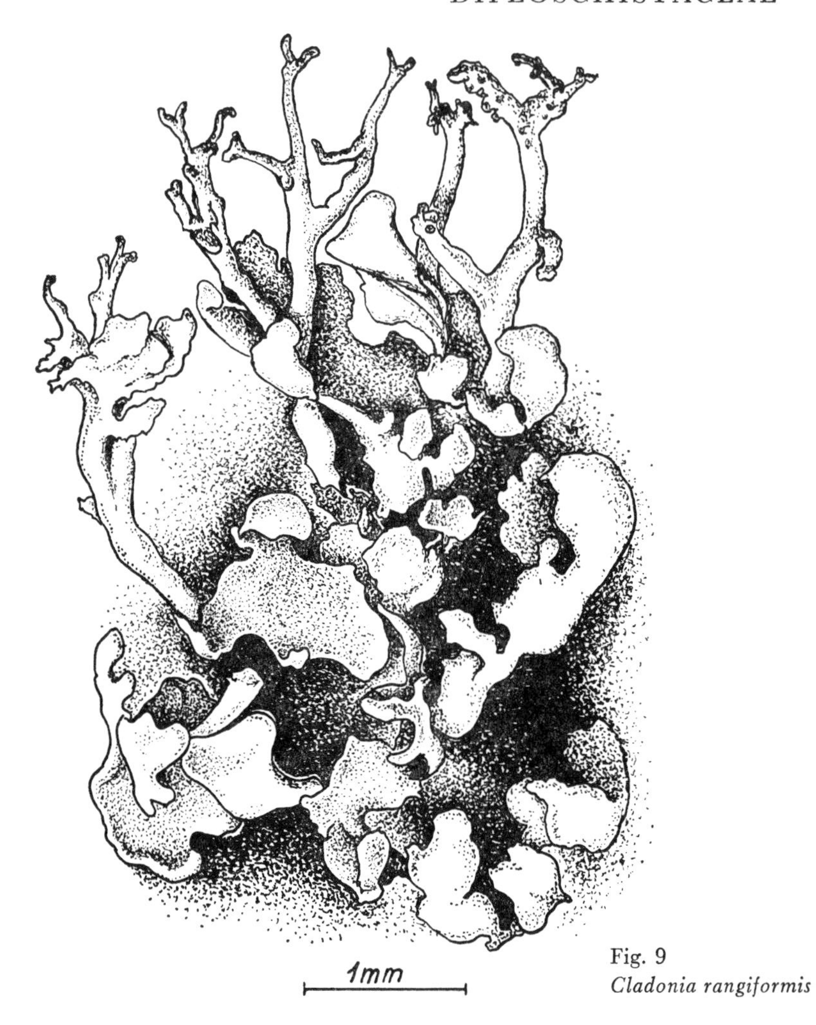

Fig. 9
Cladonia rangiformis

DIPLOSCHISTES Norm.

Thallus and apothecia as described for family. Spores muriform, colourless at first, later greenish and finally brown, often shrivelled; measurements are therefore given for green spores only.

1.	Thallus Pd+ orange. Medulla K+ red, C–. Contains norstictic acid.	**4. D. ocellatus**
–	Thallus Pd–. Medulla C+ red, K–. Contains lecanoric acid	2
2.	Spores with rounded ends	3
–	Spores with one or both ends pointed	5
3.	Thallus dark grey, rimose-areolate, *c.* 0.5 mm thick.	**1. D. actinostomus**
–	Thallus whitish, mostly continuous, more than 0.5 mm thick	4
4.	Saxicolous.	**3. D. calcareus**
–	On plant remains or on soil.	**6. D. steppicus**
5.	Saxicolous.	**5. D. scruposus**
–	On mosses, other lichens, or soil, only occasionally on rocks.	**2. D. bryophilus**

1. Diploschistes actinostomus Zahlbr., Hedwigia 31 : 34 (1892). [Plate VI : 2]

Thallus crustose, indeterminate, dark grey, thin (approx. 0.5 mm), rimose-areolate, continuous, rather even; areoles plane, irregular, 0.3–1 mm across, closely attached to the substrate. Cortical cells rather inconspicuous, colourless, roundish, nubilated by greyish crystals and covered with a transparent amorphous stratum. Algae in a continuous layer or in groups. Medulla thin, of more or less loosely entwined hyphae, obscure with greyish and brown granules and particles.

Apothecia minute, 1 or several per areole, immersed, almost spherical in section, cryptolecanorine (Dughi, 1952). Thalline margin not distinct. Disc level with the thallus, black. Exciple blackish, usually greyish-pruinose, radiate-striate in structure, colourless below and within, yellowish brown above and without. Hymenium colourless, *c.* 0.5 mm thick. Hypothecium thin, colourless except for a dark brown lower rim (= excipulum hypotheciale; Lettau, 1937). Paraphyses thin and slender. Asci with (4–6–)8 spores. Spores colourless at first, later olive-green, at maturity brown and shrivelled, ovoid or broad-ellipsoid with rounded ends, muriform with 3–7 transverse and 1–4 longitudinal septae, 25–33 × 15–22 μ.

Reactions: Thallus K–, KC–, Pd–; medulla C+ red, I+ blue in places; contains lecanoric acid (Pl. XXV : 2); inner colourless zone of exciple I+ blue; hymenium I+ yellowish.

Habitat: On basalt. Upper Galilee, Esdraelon Valley.

Distribution: In temperate regions of the Holarctis.

2. Diploschistes bryophilus (Erht.) Zahlbr., Hedwigia 31 : 34 (1892). *Lichen bryophilus* Erht., Plant Crypt. Ecsicc. No. 236 (1924). [Plate VI : 3]

Thallus indeterminate, 1–4 cm across, *c.* 1 mm thick, whitish grey, pulverulent, usually undulate and uneven, continuous, in loose contact with the substrate. Cortex colourless, very poorly developed, filled and covered with many colourless crystals. Algal layer continuous, more or less uniform. Medulla macroscopically white, composed of thin, loosely intricate hyphae, greyish-nubilated in section.

Apothecia scattered, up to 1 mm across, immersed-urceolate. Thalline margin intergrading with the surrounding thallus. Disc blackish, pruinose. Exciple 75–90 μ thick; internal hyphae colourless, perpendicular to the hymenium; externally blackish brown and 30–65 μ higher than the level of the disc (= excipulum hypertheciale; Lettau, 1937). Hymenium colourless, 100–125 μ thick. Epithecium blackish, 8–10 μ thick. Hypothecium blackish brown, 35–40 μ thick, interrupted by the colourless lower and internal zone of the exciple. Paraphyses thin and slender. Spores 4, at first colourless, later olive-green, at maturity brown and shrivelled, muriform, with 3–6 transverse and 1–3 longitudinal septae, ovoid with pointed ends, 22–36 × 12–15 μ.

Reactions: Thallus K+ yellow, KC+ yellow-brown, Pd–, B+ violet; medulla C+ red, I usually –, at some spots blue; contains diploschistic acid.

Habitat: On mossy soil and on *Cladonia pocillum* squamules in crevices of calcareous rocks, occasionally spreading on to the rock, always in the shade. Upper Galilee, Judean Mountains.

Distribution: Mediterranean and Medio-European, extending into Atlantic and Boreal territories.

3. Diploschistes calcareus Stein., Ann. Naturhist. Hofm. Wien 21 : 382 (1905). *D. tenuis* Reichert et Galun, Bull. Res. Counc. Israel 9 D: 130 (1960). [Plate VII : 4]

Thallus more or less determinate, up to 10(–15) cm across and *c*. 1 mm thick, plane and even, rarely partly wrinkled, whitish, farinose, either divided by very fine fissures into areoles 0.5–2 mm across or continuous and undivided; periphery usually more conspicuously rimose-areolate and about 2 mm thick, firmly attached to the substrate. Cortex 25–35 μ thick, nubilated by many crystals. Algae in groups or in a continuous layer; some dispersed in the medulla.

Apothecia abundant, absent from the peripheral zone, 0.5–0.7 mm across, innate. Disc black, plane. Exciple laterally 60–70 μ thick; inner part composed of colourless hyphae 1.5 μ thick, perpendicular to the hymenium; outer part composed of blackish hyphae 2.5–3 μ thick; both parts raised above the disc level to form a brush-like margin 80–120 μ high and 70–90 μ thick (= excipulum hypertheciale; Lettau, 1937). Thalline margin lacking. Hymenium colourless, 240–280 μ thick. Hypothecium 15–30 μ thick, colourless above, fuliginous below. Ascus with 4–8 spores. Spores at first colourless, brown when mature, muriform with rounded ends, with 5–7 transverse and 1–4 longitudinal septae, 28–32 × 14–17 μ.

Reactions: Thallus and medulla K–, B–; medulla C+ red; contains lecanoric acid (Pl. XXV : 2); hymenium I–.

Habitat: On calcareous rocks. Upper Galilee, Shefela, W. and C. Negev.

Distribution: Mediterranean and Irano-Turanian.

4. Diploschistes ocellatus Norm., Nyt Mag. Naturv. 7 : 232 (1853). [Plate VII : 2]

Thallus somewhat orbicular, several cm across, 1–3 mm thick, greyish with a pink tinge, more or less uniformly areolated; fissures thin but deep; areoles plane to moderately convex, 0.2–0.5 mm across, roundish to polygonal, more or less radially arranged towards the periphery. Margin sometimes thickened, bent or raised or adnate. Cortex 15–25 μ thick, colourless, clear, covered by an amorphous dark grey layer 10–20 μ thick. Algae in a continuous palisade-like layer 40–50 μ thick. Medulla colourless, nubilated.

Apothecia numerous, mainly centrally located, up to 3 mm across. Disc plane, rarely slightly concave or convex, blackish brown, pruinose. Thalline margin 70–100 μ thick, at first incurved, entire, prominent, persistent, concolourous with the thallus. Exciple 10–20 μ thick, colourless. Hypothecium colourless, 70–150 μ thick. Hymenium colourless, 90–140 μ thick. Spores 8, pale brown to brown, muriform with rounded ends, with 3–5 transverse septae and 0–2 longitudinal septae, 27–36 × 9–13 μ.

Reactions: Thallus K+ at first yellow, later rust-brown, Pd+ orange, C–; medulla K+ reddish brown; contains norstictic acid * (Pl. XXVI : 3).

* Erroneously published as containing salazinic acid (Galun, 1966).

Habitat: On calcareous substrate, usually under shade. Upper Galilee, Dan Valley, Judean Mountains.

Distribution: Mainly Mediterranean, scattered in dry enclaves of the southern Medio-European province and extending to the East as far as Medio-Asiatic territories.

5. **Diploschistes scruposus** Norm., Nyt Mag. Naturv. 7: 232 (1853). [Plate VII: 3]

Thallus spreading, dark grey, 1–3 mm thick, separated by deep, broad fissures into areoles; areoles more or less roundish, elongated or curved, sometimes contorted or plicate, in groups 1–3 mm wide, each of which has secondary complete or incomplete thin cracks; surface mealy and crumbly. Cortex colourless, composed of densely aggregated thick hyphae covered with an amorphous transparent layer. Algal cells arranged in groups. Medulla white.

Apothecia numerous, more so in the centre, usually solitary, rarely 2–3 per areole, up to 2 mm across, level with the thallus or slightly raised. Disc black, slightly pruinose, deep-concave. Thalline margin concolourous with the thallus, thick, entire or somewhat rugose, prominent or intergrading with the thallus. Exciple composed externally of thick dark brown hyphae ascending 70–80 μ above the disc (= excipulum hypertheciale; Lettau, 1937); internal hyphae colourless, perpendicular to the hymenium. Hymenium colourless, more or less square in section. Epithecium and hypothecium brown. Spores 6–8, at first colourless, later olive-green, at maturity brown and shrivelled, muriform, with 3–6 transverse septae and 1–3 longitudinal septae, one or both ends pointed, 28–35 × 10–12 μ.

Reactions: Thallus K± yellow, C+ red, KC+ red; medulla K–, C+ red, I+ blue in spots only; contains lecanoric acid (Pl. XXV: 2).

Habitat: On basalt. Upper Galilee.

Distribution: Pluriregional.

6. **Diploschistes steppicus** Reichert, Palest. J. Bot. 3: 173 (1940). [Plate VII: 1]

Thallus flat or slightly plicate, continuous or sometimes areolate-rimose, greyish white or whitish farinose, sometimes covered with minute squamules, 0.6–4 mm thick. Cortex 17–80 μ thick, opaque, dirty grey. Algal layer continuous, 30–220 μ thick. Medulla 550–3,750 μ thick, opaque, clouded.

Apothecia solitary or in groups, immersed or slightly prominent, 0.5–2 mm across. Disc black, naked or slightly pruinose, concave to plane. Thalline margin consisting of an algal layer which ends behind the exciple and never covers it. Cortex ascending alone to cover the exciple (Fig. 10). Exciple black, greyish to colourless at the base, 70–290 μ thick, raised to form an excipulum hypertheciale (Lettau, 1937) up to 230 μ high. Hypothecium blackish, 15–30 μ thick. Hymenium 120–160 μ thick, colourless. Epithecium dark brown, 5–15 μ thick. Spores 8, at maturity olive-green, finally brown and shrivelled, muriform, with 3–6 cross walls and 1–2 longitudinal walls and rounded ends, 20–30 × 9–16 μ.

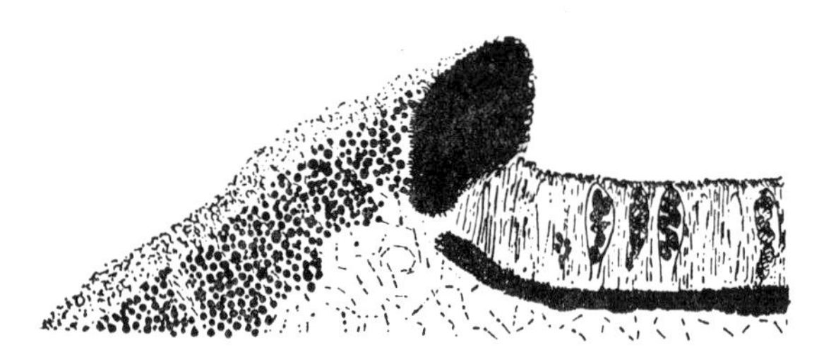

Fig. 10
Diploschistes steppicus
Section through apothecium (× 70)
(from Reichert, 1940)

Reactions: Thallus K+ yellow, C–, B+ violet; medulla K–, C+ red, I–; contains diploschistic acid.

Habitat: On remnants of *Poa, Carex* and other Gramineae species and on gypsous and calcareous soil. Judean Desert, C. and W. Negev.

Distribution: Irano-Turanian.

PERTUSARIACEAE

Thallus crustose, effuse or determinate, firmly attached to the substrate by medullary hyphae; corticate or non-corticate above. Phycobiont belonging to the Chlorophyceae. Apothecia one or several, immersed in thalline warts. Asci with 1–8 spores. Spores large, thick-walled, usually colourless, simple or 1-septate.

Pertusaria DC.

Thallus epi- or hypophloeodal or epilithic, with or without isidia and / or soredia, stratified, corticate or non-corticate above. Apothecia generally immersed in thalline warts with a poriform or discoid opening. Paraphyses slender, branched and intricate. Spores simple, colourless, large, thick-walled.

1. Thalline warts sorediate, K+ yellow→red, Pd+ orange. Asci with one spore. **4. P. multipuncta** var. **leptosporoides**
– Thalline warts not sorediate, K– or slightly yellow, Pd–. Asci with more than one spore 2
2. Thallus K–. Asci with 2 spores 3
– Thallus K± yellow. Asci usually with 4 spores. **3. P. leucostoma** var. **areolascens**
3. Spores less than 100 μ long. **1. P. carmeli**
– Spores more than 100 μ long. **2. P. ilicicola**

1. Pertusaria carmeli Reichert et Galun, Israel J. Bot. 14 : 11 (1965). [Fig. 11]

Thallus epiphloeodal, cream-coloured, very thin, effuse to verruculose, mostly consisting of contiguous fertile warts with a broadened base forming a more or less continuous crust.

Apothecia usually solitary, rarely 2–3 per wart; warts with openings at first poriform, later discoid. Disc plane, up to 0.5 mm across, covered with a thick whitish pruina. Hymenium obscured by many small crystals. Epithecium granulose, greyish. Hypothecium colourless. Paraphyses ramose, loosely entwined. Spores 2, simple, ellipsoid, 65–95 × 25–27 μ; cell wall 3–4.5 μ thick; cytoplasm forming a hyaline network.

Reactions: Thallus K–, C–; hymenium I+ at first blue, later dark violet.

Habitat: On *Quercus calliprinos*. Mt. Carmel.

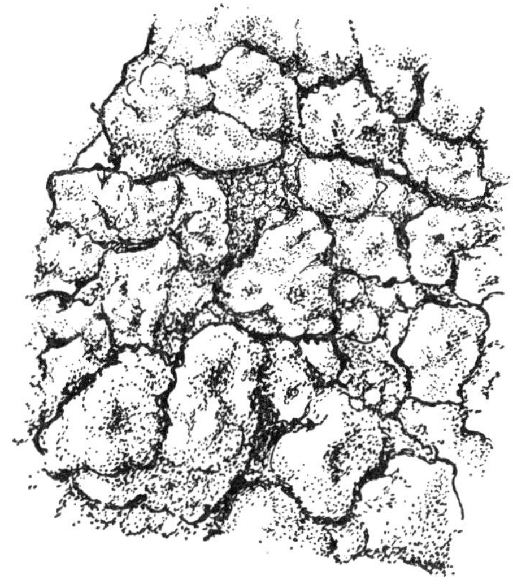

Fig. 11
Pertusaria carmeli

2. Pertusaria ilicicola Harm., Lich. de France 5: 1120 (1913). [Plate VIII: 1]

Thallus epiphloeodal, indeterminate, whitish with a yellowish tinge, 300–600 μ thick, minutely granulose-verrucose and irregularly rimose.

Apothecia one or several to each wart; warts with openings at first poriform, later discoid. Disc plane or nearly so, up to 1 mm across, blackish, almost always white-pruinose, somewhat immersed in the warts and surrounded by an irregular, densely pruinose margin. Hymenium 150–250 μ thick, colourless. Epithecium brown. Hypothecium colourless. Paraphyses few, branched and intricate above. Spores 2, colourless, ellipsoid, 112–135 × 60–70 μ, spore wall 4–8 μ thick.

Reactions: Thallus K–, C–, Pd–; hymenial gelatine I–; asci I+ blue; spores I–.

Habitat: On branches of *Rhamnus palaestina* and *Quercus calliprinos*. Upper Galilee.

Distribution: Mediterranean.

3. Pertusaria leucostoma (Bernh.) Mass. var. **areolascens** Erichs., in Rabh. Krypt.-Fl. 9(5): 430 (1935). [Plate VIII: 2]

Thallus epiphloeodal, indeterminate white to greyish white, rather thick, at first irregularly rimose-areolate, later fine-verrucose.

Apothecia globose, usually solitary, sometimes 2–3 per wart; warts scattered or in groups, *c.* 1 mm across, with a constricted base, tumid, convergent margin and poriform opening. Hymenium and hypothecium colourless. Paraphyses branching in a network. Asci usually with 4 spores, sometimes with 3 or 5. Spores colourless, ellipsoid, 76.5–102 × 23–36 μ, spore wall 2–6 μ thick.

Reactions: K± slightly yellow, C–, KC–, Pd–; asci and spores I+ blue; hymenial gelatine I–.

Habitat: On bark of *Quercus calliprinos*. Judean Mountains.

Distribution: Reported only from Corfu.

4. **Pertusaria multipuncta** (Turn.) Nyl. var. **leptosporoides** Erichs., in Rabh. Krypt.-Fl. 9(5): 611 (1936).

Thallus epiphloeodal, expanding over large areas, rather thin, at first whitish, later with a brownish tinge, sometimes rather smooth, most often rough, warty and wrinkled.

Apothecia usually solitary, closed in warts at first and later opening up slightly, finally turning into a soredial heap. Disc black, pruinose, plane or somewhat concave, 0.3–0.5 mm across. Hymenium colourless, 140–160 μ thick. Epithecium thick, olive-brown. Paraphyses somewhat branched and loosely entwined. Ascus with one spore. Spores colourless, ellipsoid, 115–130 × 50–56 μ, spore wall 5–8 μ thick.

Reactions: Thallus K+ yellow, C+ yellow, Pd+ orange; medulla K+ yellow→ red, I–; contains norstictic acid * (Pl. XXVI: 3); asci I+ blue.

Habitat: Common on *Quercus calliprinos*. Upper Galilee.

Distribution: Reported from Bohemia.

ACAROSPORACEAE

Thallus crustose, squamulose to subfoliose or obsolete, stratified or rarely unstratified. Algae belonging to the Chlorophyceae. Apothecia immersed or superficial, with an exciple or a thalline margin, sometimes with both. Asci with many spores. Spores colourless, minute, with thin walls, simple (exceptionally 2-celled).

1. Apothecia with a thalline margin. **1. Acarospora**
– Apothecia without a thalline margin. **2. Sarcogyne**

1. ACAROSPORA Mass.

Thallus crustose or areolate-squamulose, uniform or marginally lobed, distinctly paraplectenchymatous-corticate, attached to the substrate by hyphae, rhizinae lacking.

* Erroneously published as containing salazinic acid (Galun & Lavee, 1966).

Apothecia lecanorine, immersed or becoming superficial, solitary or several per areole (or squamule). Hypothecium colourless or yellowish, subtended by an algal layer. Spores numerous, simple, mostly small, colourless, ellipsoid or spherical.

1. Saxicolous 2
- Terricolous 3
2. Thallus dark brown, areolate, sublobate at the circumference. **1. A. bornmuelleri**
- Thallus whitish, areolate, indeterminate. **2. A. areolata**
3. Thallus ochroid or sand-coloured, areolate-squamulose, indeterminate. **3. A. murorum**
- Thallus a chalk-white continuous effigurate crust. **4. A. reagens** f. **radicans**

1. Acarospora bornmuelleri Stein., Ann. k. k. Naturhist. Hofmus., Wien 30 : 29 (1916) and Ann. Naturhist. Staatsmus., Wien 34 : 15 (1921). [Plate IX : 4]

Thallus crustose, dark brown, areolate; areoles plane and even, angular, *c.* 1 mm across, 2–3 mm thick; circumference sublobate, each lobe showing laterally an uppermost hard and blackish brown layer *c.* 0.25 mm thick, subtended by a softer brown to reddish brown layer 1–1.5 mm thick, the rest belonging to the milk-white medulla. Cortex 40–60 μ thick, colourless except for the brown pigmented cell walls of the uppermost cell rows, covered by a colourless amorphous stratum 5–8 μ thick, each areole with a lateral dark brown pseudocortex corresponding to the depth of the separating fissures, the areoles being connected by the lower part of the medulla. Algal layer *c.* 100 μ thick, continuous or in groups separated by strands of parallel hyphae. Medulla nubilated by many crystals and substrate particles.

Apothecia rare to rather numerous, usually solitary, sometimes 2–3 to each areole, both marginal and central areoles fertile, up to 1 mm across, usually less. Disc level with the thallus, plane, granular, concolourous with the thallus or darker. Hymenium colourless, 100–120 μ thick. Epithecium brown. Hypothecium colourless. Exciple 15–20 μ thick, composed of parallel radiating hypothecial hyphae, colourless except for the brown-pigmented exterior cell walls. Paraphyses coherent, septate, with brown apical cells. Spores numerous, colourless, subglobose or globose, 3–5 μ across.

Reactions: Thallus and medulla K–; cortex C+ orange-red (a very thin line of colouration is formed at the upper part of the cortex below the brown region); no chemical constituents detected.

Habitat: On basalt, in open and dry positions. Upper Jordan Valley, Lower Galilee.

Distribution: Mesopotamian, Irano-Anatolian.

2. Acarospora areolata Reichert et Galun, Bull. Res. Council Israel, Bot. 9D: 132 (1960). [Plate IX : 2]

Thallus areolate; areoles (0.5–)1–2.5 mm across and 0.4–0.8 mm thick, dispersed, in small groups or contiguous, greyish white, densely pruinose, rough and granulous, slightly convex, the larger areoles occasionally with thin splits; lower side pale, firmly attached to the substrate. Cortex *c.* 30 μ thick, translucent except for an exterior brown amorphous stratum *c.* 15 μ thick; pruina 10–20 μ thick. Algal layer 150–300 μ thick, interrupted by medullary hyphae perpendicular to the surface. Medulla greyish, made opaque by masses of substrate particles.

Apothecia numerous, mostly solitary, sometimes 2–3 to each areole. Disc more or less level with the thallus, 0.5–1.5 mm across, black, dull, plane. Thalline margin concolourous with the thallus. Exciple colourless, 15–40 μ thick at the base and expanding to 150 μ above. Hymenium colourless, 100–150 μ thick. Epithecium yellowish brown. Hypothecium colourless, grumous. Paraphyses septate; apices slightly swollen and yellowish brown, conglutinate. Asci with *c.* 100 spores. Spores colourless, cylindrical, 4.5–6 × 1.5–2 μ.

Reactions: Thallus and medulla K–, C–, Pd–; hymenium I+ blue; no chemical constituents detected.

Habitat: On calcareous stones. Rare. Judean Desert, C. Negev.

3. Acarospora murorum Mass., Mem. Lichgr. 130 (1853). [Plate IX: 1]

Thallus areolate-squamulose, indeterminate, more or less concrescent into an areolate crust, ochroid to sand-coloured; squamules plane, 1–2 mm across, more or less intergrading with the substrate. Cortex translucent, 25–85 μ thick, sometimes with an uppermost yellowish cover 10–15 μ thick and covered by an amorphous layer 60–100 μ thick. Algal layer 40–70 μ thick, irregularly interrupted by perpendicular strands of medullary hyphae. Medulla nubilated by granules and substrate particles; lower part of medulla brown.

Apothecia numerous, minute, solitary or 2–3 to each squamule. Disc dark brown, epruinose, concave or plane, on a level with the thallus. Thalline margin concolourous with the thallus, slightly prominent. Hypothecium colourless, cellular, indistinctly separated from the colourless hymenium, 115–160 μ thick. Epithecium yellowish brown, covered by an amorphous layer 20–40 μ thick. Paraphyses coherent, septate, with slightly broadened apical cells; spores very numerous, colourless, subglobose, 6.5–8.5 × 4–5 μ, usually with one large vacuole.

Reactions: Hymenium I+ blue; no chemical constituents detected.

Habitat: On fine sandy soil in shade. C. Negev.

Distribution: Mediterranean.

4. Acarospora reagens Zahlbr. f. **radicans** (Nyl.) Magn., Kungl. Svensk. Vetensk.-Akad. Handl. 4: 272 (1929). *Lecanora schleicheri dealbata* f. *radicans* Nyl., Lich. in Aegypto a Ehrenb. coll. 63 (1864). [Plate IX: 3]

Thallus a more or less continuous, chalk-white, hard, effigurate crust *c.* 1 mm thick, 2–5 cm across; surface occasionally cracked, covered with a thick whitish pruina. Cortex uneven, 15–30(–45) μ thick, opaque, delimited by a brown line of the uppermost pigmented cells; pruina 45–80 μ thick. Algal layer 50–120 μ thick, continuous. Medulla nubilated with crystals and substrate particles, connected to the substrate by thick root-like medullary hyphae.

Apothecia numerous, up to 1.2 mm across. Disc plane or slightly convex, more or less level with the thallus, black, naked. Thalline margin concolourous with the thallus, thin and evanescent. Exciple 15–20 μ thick, laterally not continued above or below; interior dark brown; exterior black. Hymenium 170–180 μ thick, colourless except for the gelatinous brown epithecium. Hypothecium colourless, inspersed and bordered below by groups of algal cells. Paraphyses simple, septate, with a broadened brown apical cell. Spores numerous, colourless, spherical, 3–4.5 μ across.

Reactions: Thallus K+ yellow→red, C−; medulla Pd+ orange yellow; contains norstictic acid (Pl. XXVI: 3); hymenium I+ blue.

Habitat: On fine soil and on loess. Judean Desert, C. and W. Negev.

Distribution: Previously reported from Bir-Haman, a desert location in Egypt (Nylander, 1864) and from the Judean Desert (Müller Argovensis, 1884).

2. SARCOGYNE Flotow

Thallus crustose, sometimes marginally lobed, or often poorly developed to obsolete, corticate above or undifferentiated, attached to the substrate by hyphae; rhizinae lacking. Apothecia biatorine or lecideine. Hypothecium colourless or dark. Spores numerous, simple, colourless, small, spherical to oblong or ellipsoid.

1. Sarcogyne pruinosa (Sm.) Körb., Syst. Lich. Germ., 267 (1855). *Lichen pruinosus* Sm., Engl. Bot. 1: 2244 (1811).

Thallus very thin, whitish grey, farinose, inconspicuous. Phycobiont *Myrmecia biatorellae* (Geitler, 1960).

Apothecia abundant, partly scattered, partly grouped, circular, 0.5–1.5 mm across, at first immersed, later sessile and appressed, leaving a shallow pit when removed. Disc black, dark brown when wet, plane, pruinose. Exciple entire, persistent, blackish, naked, up to 100 μ thick laterally; exterior dark brown, gradually becoming colourless towards the interior, which is composed of thin radiating hyphae. Hymenium 80–120 μ thick, colourless. Epithecium yellowish brown to brown. Hypothecium colourless, inspersed by colourless crystals. Paraphyses conglutinate, simple, septate, slightly broadened at the apex. Spores 100–200, simple, colourless, oblong to almost cylindrical, 3–5 × 1–2 μ.

Reactions: Hymenium I+ blue.

Habitat: On calcareous rocks. Upper Galilee.

Distribution: Very common in the N. Hemisphere.

LECANORACEAE

Thallus crustose, uniform, lobate or squamulose, stratified, corticate above or non-corticate, attached to the substrate by medullary hyphae. Phycobiont belonging to the Chlorophyceae. Apothecia innate or superficial, lecanorine. Exciple present or lacking. Paraphyses simple or branched and intricate. Asci usually with 8 spores, sometimes less, and in some cases up to 32. Spores colourless, single-celled or septate.

1. Spores simple 2
- Spores septate 4
2. Spores more than 30 μ long. **3. Ochrolechia**
- Spores less than 30 μ long 3
3. Thallus of broad and thick lobes or squamules. Cortex thick and uniform. **5. Squamarina**
- Thallus crustose or lobate. Cortex indistinct or thin. **2. Lecanora**
4. Thallus crustose, uniform. **1. Lecania**
- Thallus lobate-effigurate. **4. Solenopsora**

1. Lecania Mass.

Thallus crustose, uniform, stratified. Upper cortex differentiated or not. Apothecia small, lecanorine. Exciple present or absent. Spores usually 8, colourless, 2–4-celled, thin-walled.

1. Spores 1-septate 2
- Spores 3-septate 3
2. Thallus fairly well developed. Apothecia at first innate, later adnate. **1. L. erysibe**
- Thallus inconspicuous. Apothecia superficial from the beginning. **4. L. subcaesia**
3. Corticolous. **2. L. koerberiana**
- Saxicolous. **3. L. nylanderiana**

1. Lecania erysibe Mudd, Man. Brit. Lich. 141 (1861). [Plate VIII : 3]

Thallus crustose, thin, indeterminate or nearly orbicular, areolate, dark greyish brown, 4–5 cm across.

Apothecia numerous, occupying almost all the areoles, 0.4–1 mm across, at first immersed, later superficial. Disc dark brown to blackish, naked or slightly pruinose, at first plane, later convex. Thalline margin thin, smooth or nearly so, concolourous with the thallus, disappearing rather soon. Exciple persistent, colourless, fan-shaped. Hymenium colourless. Epithecium *c.* 20 μ thick, brownish, gelatinous. Hypothecium colourless. Paraphyses septate, simple, conglutinate, with broadened tips embedded in a brownish gelatinous stratum. Spores 8, colourless, ellipsoid, 2-celled, septum very thin, 11–15(–17) × 3.5–5 μ.

Reactions: Thallus K–, C–, KC–.

Habitat: On calcareous rocks and on cement walls. Upper Galilee.

Distribution: Rather common in the N. Hemisphere.

2. Lecania koerberiana Lahm., in Körb. Parerg. Lich. 68 (1859).

Thallus very thin, granulose, inconspicuous, greenish grey.

Apothecia numerous, scattered, orbicular, minute, 0.2–0.5 mm across, at first immersed, later sessile. Disc at first plane, soon becoming convex, dark brown or almost black, dull, epruinose. Thalline margin somewhat paler than the disc, thin, entire, soon disappearing. Hymenium colourless. Epithecium thin, brownish. Hypothecium pale yellowish, subtended by a continuous algal layer. Paraphyses coherent and capitate.

Spores 8, colourless, 4-celled, somewhat curved and constricted at the thin septae, 14–18 × 3–6 μ.

Reactions: K–, C–, KC–.

Habitat: On young twigs of olive trees. Judean Mountains.

Distribution: Mediterranean and Medio-European.

3. **Lecania nyderlandiana** Mass., Sched. Grit. Lich. Ital. 152 (1855).

Thallus effuse, irregularly cracked-areolate to coarsely granulose, obscurely greyish or sordid whitish.

Apothecia rather numerous, 0.5–0.8 mm across. Disc at first plane, later convex, brown or blackish, usually pruinose. Thalline margin thin, whitish grey, soon disappearing. Hymenium and hypothecium colourless. Epithecium brownish and gelatinous. Spores 8, subfusiform, sometimes curved, colourless, 4-celled, 15–18 × 4–5 μ.

Reactions: Thallus K–, C–; hymenium at first blue, later turning wine-red.

Habitat: On soft calcareous rocks and cement walls. Upper Galilee, Mt. Carmel.

Distribution: Euro-Siberian and Mediterranean; also found in N. America.

4. **Lecania subcaesia** (Nyl.) Szat., Ann. Mus. Nat. Hung. 8: 137 (1957). *Lecanora subcaesia* Nyl., Flora 64: 538 (1881). [Plate VIII: 4]

Thallus inconspicuous.

Apothecia numerous, scattered, loosely attached to the substrate by short stalks, up to 1 mm across. Disc black, densly caesio-pruinose, at first plane and surrounded by a whitish thalline margin, later becoming convex and the margin becoming retrorse. Hymenium colourless, 50–60 μ thick. Epithecium fuliginous, gelatinous. Hypothecium colourless, densely aggregated, 130–150 μ thick. Algal layer interrupted in the centre. Paraphyses usually unbranched, septate, the two apical cells broader and shorter than the rest. Spores 8, colourless, 2-celled with a thin septum, 10–12 × 3–3.5 μ.

Reactions: Hymenium I+ lower part blue, upper part greenish; hypothecium I–.

Habitat: On calcareous rocks and stones. C. Negev.

Distribution: Mediterranean and Irano-Turanian.

2. LECANORA Ach.

Thallus crustose, uniform, determinate or lobate-effigurate, stratified, attached to the substrate by medullary hyphae. Phycobiont *Trebouxia* (Ahmadjian, 1967) (Fig. 1A). Apothecia innate or adnate, lecanorine, with thalline margin only, or with thalline margin and exciple. Paraphyses simple or branched. Asci with 8 spores, sometimes less. Spores simple, colourless, spherical, ovoid or ellipsoid.

Subgen. ASPICILIA Mass. Thallus crustose, effuse or somewhat determinate. Apothecia innate (at least at first), disc concave or plane.

1.	Thallus 1–3 mm thick.	**1. L. desertorum**
–	Thallus less than 1 mm thick	2
2.	Medulla K+ deep yellow→orange-red.	**2. L. microspora**

- Medulla K– 3
3. Growing on flint. **3. L. hoffmannii**
- Growing on a calcareous substrate 4
4. Thallus inconspicuous. Apothecia adnate at maturity. **4. L. contorta**
- Thallus in distinct layers. Apothecia remaining innate. **5. L. farinosa**

1. Lecanora desertorum Kphbr., Verh. Zool.- Bot. Ges. Wien 17 : 601 (1867). [Plate X : 1]

Thallus attached to the substrate, very often in the form of a wide zone around the edge of small stones, sometimes covering the entire upper surface and part of the lower surface of small pieces of gravel, some very small stones (1–1.5 cm) or earth clumps entirely overgrown on both sides; thallus sand-coloured (almost concolourous with the surrounding loess), dull, rough, slightly farinose in places, verrucose-areolate; areoles congested, at first plane, later prismatic to subglobose, 1–3 mm thick. Hypothallus concolourous with the thallus or somewhat darker, especially well developed on flint.

Apothecia in almost every areole, usually solitary, rarely 2–3. Disc appearing as a narrow opening because of the very thick, elevated and tumescent thalline margin. Ascocarps very much like perithecia, 800–1000 μ across. Thalline margin 330–350 μ thick, lateral to the base of the hymenium and expanding below. Cortex colourless, clear, paraplectenchymatous, 80–250 μ thick, covered by a more or less uniform layer of crystals, 30–35 μ thick. Algae in groups in the medullary tissue. Disc blackish, pruinose, plane. Hymenium colourless, 160–180 μ thick. Epithecium greenish, gelatinous, covered by a layer of grey granules. Hypothecium colourless, more or less uniform, 60–65 μ thick. Algal layer lacking below the hypothecium, but with a few groups of algal cells dispersed here and there. Spores 4, colourless, globose, 18–23 μ.

Reactions: Thallus and medulla K–, C–, Pd–; hymenium and hypothecium I+ blue; no chemical constituents detected.

Habitat: On flinty gravel, on small calcareous stones or earth clumps, and sometimes on pieces of pottery. Judean Desert, W. Negev.

Distribution: Irano-Turanian; extremely xerophilic.

2. Lecanora microspora Zahlbr., Oester. Bot. Z. 53 : 241 (1903).

Thallus determinate, thin and very firmly attached to the substrate, 3–5 cm across, finely rimose-areolate; areoles plane and even, ash-greyish with a very fine rufous rim which gives the thallus a reddish brown tinge in places; the narrow peripheral zone radiate-lobate; lobules about 1 mm long, confluent. Hypothallus dark grey.

Apothecia abundant, innate, 1–3, on almost every areole except near the periphery, 0.3–0.6 mm across. Disc plane to concave, black, naked or slightly pruinose. Hymenium and hypothecium colourless. Epithecium gelatinous, olive-green. Spores 8, oval-ellipsoid, simple, 14–16 × 9–10 μ.

Reactions: Thallus K± yellow; medulla K+, at first deep yellow, later orange-red; hymenium I+ greenish; hypothecium I+ blue.

Habitat: On calcareous rocks. Upper and Lower Galilee, Mt. Carmel.

Distribution: Mediterranian, Irano-Turanian.

3. Lecanora hoffmannii Müll. Arg., Flora 72 : 511 (1889). [Plate X : 2]

Thallus indeterminate, buff-grey, areolate, 2–5 cm across, 250–510 μ thick with an amorphous cover 15–40 μ thick; areoles angular, more or less uniform, 0.5–0.8 (–1.0) mm across, plane and even, only the fertile areoles moderately elevated. Hypothallus sand-coloured, 1–3 mm broad.

Apothecia abundant, usually single, sometimes 2–3 to each areole, 0.2–0.5 mm across, shapeless, plane, black, pruinose. Exciple 15–20 μ thick at the base, expanding to 70 μ above. Hypothecium colourless, 30–40 μ thick. Epithecium gelatinous, olive-green. Spores usually 4, subglobose, 18–21 × 15–20 μ.

Reactions : Thallus and medulla C–, K–, Pd–; no chemical constituents detected.

Habitat : On flint in hammada deserts and loess plains. C. and W. Negev.

Distribution : Mainly Irano-Turanian and Saharo-Arabian; also extending into the Mediterranian region.

4. Lecanora contorta (Hoffm.) Stein., Verh. Zool.-Bot. Ges. Wien 65 : 199 (1915). *Lichen rupicola* Hoffm., Enum. Lich. 23 (1784).

Thallus very thin, whitish, granulose-powdery, continuous, indeterminate. Hypothallus absent.

Apothecia numerous, dispersed or in groups, 0.2–0.8 mm across, at first innate, later adnate. Disc plane, black with a thick white pruina. Thalline margin thick, prominent, heavily pruinose, persistent. Hymenium and hypothecium colourless, with no distinct limit between them. Epithecium olive-green, gelatinous. Spores 4, spherical, colourless, 18–26 μ across.

Reactions : Thallus K–, C–, KC–.

Habitat : On limestone. Mt. Carmel.

Distribution : Common in the N. Hemisphere.

5. Lecanora farinosa Nyl., Bull. Soc. Linn. Normand., 2 / 6 : 307 (1872). [Plate X : 4]

Thallus effuse over several centimetres, whitish-farinose, irregularly rimose in places, especially towards the periphery where it sometimes becomes yellowish grey, 0.5–0.8 mm thick, covered by an amorphous layer 30–50 μ thick.

Apothecia numerous, minute, innate, solitary or 2–3 together in a wart-like elevation. Disc 300–800 (–1000) μ across, 180–250 μ deep, plane, orbicular or shapeless, black, usually pruinose. Hypothecium 15–20 μ thick, colourless. Epithecium gelatinous, olive-green. Spores 1–4, colourless, subglobose, 21–23 × 18 μ.

Reactions : Thallus and medulla K–, C–, Pd–; no chemical constituents detected.

Habitat : Common on calcareous rocks and stones in hammada deserts and loess plains. Judean Desert, W. and C. Negev.

Distribution : Mediterranean, Irano-Turanian and Saharo-Arabian.

Subgen. EULECANORA Th. Fr. Thallus crustose, uniform or determinate; apothecia superficial.

1. Corticolous 2

- Saxicolous 5
2. Thallus K–. Apothecia pruinose. **9. L. hageni**
- Thallus K+ yellow. Apothecia epruinose 3
3. Thalline margin more or less uniformly crenate. Hymenium I+ reddish brown. **13. L. subrugosa**
- Thalline margin entire, rarely crenulate. Hymenium I+ blue 4
4. Thallus verrucose, greenish grey. **11. L. scrupulosa**
- Thallus granulose-subareolate, creamy grey. **10. L. olea**
5. Growing on calcareous rocks 6
- Growing on basalt 7
6. Thallus inconspicuous. Hymenium and hypothecium colourless. **7. L. crenulata**
- Thallus well developed. Hymenium violet; hypothecium yellowish. **6. L. atra**
7. Thallus rimose-areolate, C+ orange-red (olivetoric acid). **12. L. subplanata**
- Thallus coarsely verrucose, C± yellowish (gangaleoidin). **8. L. gangaleoides**

6. Lecanora atra Ach., Lichgr. Univ. 344 (1810). [Plate X : 3]

Thallus determinate, 1–5 cm across, greyish white with a yellowish tinge (turning more yellowish with time), uneven, granulose-verrucose to rugose, *c.* 2 mm thick in the centre, *c.* 1 mm thick towards the periphery. Hypothallus greyish black. Cortex indistinct. Algal layer almost reaching the outermost colourless, amorphous cover.

Apothecia numerous, dispersed or approaching, sessile, orbicular or slightly deformed when mature. Disc black, smooth, epruinose, plane. Thalline margin *c.* 150 μ thick, entire or crenulate, persistent, prominent, concolourous with the thallus. Hymenium violet, 70–80 μ thick. Subhymenium colourless, 70–80 μ thick. Epithecium fuliginous, *c.* 25 μ thick. Hypothecium yellowish, 140–160 μ thick, subtended by a continuous algal layer. Paraphyses embedded in a thick, gelatinous matrix which is violet above. Spores 8, simple, colourless, ovoid or ellipsoid, 11–15 × 5–6 μ.

Reactions: Thallus C–, K+ yellow; contains atranorin (Pl. XXV : 1).

Habitat: Frequent on calcareous rocks. Upper and Lower Galilee, Mt. Gilboa, Mt. Carmel, Judean Mountains, C. Negev.

Distribution: Cosmopolitan.

7. Lecanora crenulata (Dicks.) Hook., in Sm. Engl. Fl. 5 : 194 (1844). *Lichen crenulatus* Dicks., Fasc. Pl. Cryptog. Brit. Pl. 2 : 709 (1776). [Plate X : 5]

Thallus very thin, effuse, greyish white, often scarcely visible.

Apothecia numerous, scattered or crowded, usually less than 1 mm across, at first appearing as small wart-like outgrowths with a minute pore. Disc plane when mature, dark brown or black, brown when wet, pruinose. Thalline margin 90–100 μ thick, white, prominent, becoming deeply and more or less regularly crenulate. Exciple colourless, 10–30 μ thick. Epithecium brown, granulose. Hypothecium colourless, 60–80 μ thick, subtended by an interrupted algal layer. Paraphyses capitate, slightly constricted at the septae. Spores 8, ellipsoid, 10–15 × 4.5–6 μ.

Reactions: K–, C–.

Habitat: On calcareous rocks. W. Negev.

Distribution: Euro-Siberian and Mediterranean, slightly extending into the Irano-Turanian region.

8. Lecanora gangaleoides Nyl., Flora 55 : 354 (1872). [Plate X : 6]

Thallus spreading, somewhat effigurate, thick, coarsely warted-areolate, gradually thinner towards the periphery, ash-grey. Hypothallus usually distinct, whitish.

Apothecia numerous, more or less equally dispersed, at first immersed but soon sessile, up to 1 mm across. Disc plane, black, epruinose. Thalline margin *c.* 200 μ thick, smooth, prominent, entire, persistent, concolourous with the thallus. Exciple colourless, 10–30 μ thick. Hymenium and hypothecium colourless. Epithecium green. Paraphyses with slightly clavate and coloured tips. Spores 8, simple, ellipsoid, 12.5–15 × 6.8–7.6 μ.

Reactions: Thallus K+ yellow, C± yellow; contains atranorin (Pl. XXV : 1) and gangaleoidin; hymenium I+ blue; hypothecium (upper part) I+ violet.

Habitat: On basalt. Upper Galilee.

Distribution: Medio-European, Atlantic and Mediterranean.

9. Lecanora hageni Ach., Lichgr. Univ. 367 (1810).

Thallus effuse, very thin, minutely granular, greyish white, scarcely visible.

Apothecia numerous, dispersed or crowded , 0.3–0.5 mm across, adnate. Disc plane, dull brown with a blue-grey pruina. Thalline margin thin, white, persistent, slightly prominent, entire or crenulate. Epithecium granulose, brownish. Hypothecium colourless. Spores 8, simple, colourless, ellipsoid, 10–16 × 4–6.5 μ.

Reactions: Thallus K–, C–, KC–.

Habitat: On smooth bark and young twigs. Upper Galilee, Esdraelon Valley, Judean Mountains.

Distribution: Throughout Europe and N. America.

10. Lecanora olea Reichert et Galun, Bull. Res. Counc. Israel 6 D(4): 238 (1958). [Plate XII : 1]

Thallus thin, granulose-subareolate, creamy greyish; hypothallus somewhat paler.

Apothecia abundant, densely crowded, often deformed by pressure, sessile, 0.5–1.2 mm across. Disc plane or somewhat convex, reddish brown, epruinose. Thalline margin entire, 75–100 μ thick, centrally nubilated by crystals, surrounded by algae. Cortex yellowish brown, hyphose, 40–50 μ at the base, 20–30 μ lateral to the hymenium, 10–15 μ at the edge, occasionally somewhat flexuous, concolourous with the thallus. Exciple 0–30 μ thick, colourless, paraphysoid. Hymenium colourless, 70–100 μ thick. Epithecium brown, gelatinous, *c.* 15 μ thick. Hypothecium colourless, 45–75 μ thick, subtended by a continuous algal layer. Spores 8, simple, colourless, ovoid with many vacuoles, 12–15 × 6–7.5 μ.

Reactions: Thallus K+ yellow, C–, Pd–; contains usnic acid (Pl. XXV : 4); hymenium I+ blue.

Habitat: Common on olive trees, *Ceratonia siliqua* and *Quercus calliprinos*. Esdraelon Valley, Mt. Carmel, Shefela, Judean Mountains.

11. Lecanora scrupulosa Ach., Lichgr. Univ. 375 (1810).

Thallus indeterminate, grey to greyish green, verrucose, almost all verrucae finally developing apothecia.

Apothecia deformed by compression. Disc at first poriform, later plane, up to 1 mm across, usually less, reddish brown or pale brown. Thalline margin 80–150 μ thick, prominent, concolourous with the thallus, at first entire, later flexuous and sometimes crenulate. Exciple hyphose, radiate, 30–40 μ thick. Hymenium yellowish below, colourless above; epithecium brownish, granulose. Hypothecium colourless. Paraphyses of uniform width, septate and branched. Spores 8, simple, colourless, ellipsoid, 12–15 × 7–9 μ.

Reactions: Thallus K+ yellow, Pd–, C–; hymenium I+ blue.

Habitat: On *Quercus calliprinos*. Mt. Carmel.

Distribution: C. and W. Europe and some localities in the Mediterranean region. Rare.

12. Lecanora subplanata Nyl., Flora 64 : 350 (1881). [Plate XI : 1]

Thallus somewhat orbicular and 1–4 cm across when single, pale greyish green or occasionally pale grey, farinose, rimose-areolate; areoles more or less plane, 0.25–0.5 mm across, *c*. 0.5 mm thick; peripheral areoles usually whitish, bigger and bullate, and forming a black margin when in contact with certain other species; several thalli usually confluent.

Apothecia numerous, scattered or crowded, innate to adnate, up to 1.5 mm across. Disc black or obscure, at first plane, later convex; some naked, most with dense white pruina. Thalline margin thin, pruinose, finally disappearing. Cortex of perpendicular colourless hyphae nubilated by greyish crystals. Algal layer limiting the hymenium. Exciple lacking or very thin. Epithecium colourless or fuliginous, covered by the pruina granules. Hymenium and hypothecium colourless. Paraphyses septate, simple; apical cell clavate with a fuliginous cell wall. Spores 8, simple, colourless, ovoid-ellipsoid, some slightly constricted at the centre, often with two vacuoles (giving the impression of 2-celled spores), 8.5–10 × 3.5–5.5 μ.

Reactions: Thallus K+ yellow, C+ orange-red; contains olivetoric acid (Pl. XXIV : 5) and usnic acid (Pl. XXV : 4); hymenium I+ blue.

Habitat: Frequent on basalt. Upper Galilee, Esdraelon Valley.

Distribution: Mediterranean and Medio-European.

13. Lecanora subrugosa Nyl., Flora 58 : 15 (1875).

Thallus indeterminate, coarsely warty-granulose, greyish white to almost white. Hypothallus whitish.

Apothecia numerous and crowded, 0.5–1.5 mm across, urn-shaped. Disc dark reddish brown, plane, epruinose. Thalline margin elevated, persistent, distinctly crenate, whitish. Cortex hyphose, gelatinous-yellowish, 10–20 μ thick above, 30–40 μ thick at the base; algal layer distinct between the nubilated medulla and the cortex, reaching almost the edge of the margin. Hymenium 60–80 μ thick, gelatinous, reddish brown above, colourless below; covered by a transparent amorphous layer. Hypothecium colourless. Spores 8, simple, colourless, ellipsoid, 10–16 × 5–7 μ.

Reactions: Thallus K+ yellow, KC−, C−, Pd−; hymenium I+ reddish brown.
Habitat: On olive trees. Judean Mountains.

Subgen. PLACODIUM (Pers.) Poelt. Thallus effigurate-lobate or subfoliose, cortex encrusted with greyish yellow granules. Apothecia superficial, sometimes immersed at first.

1. Thallus whitish or grey, pruinose 2
- Thallus yellowish green, epruinose (or rarely with a thin pruina along the margins only) 3
2. Thallus K+ yellow turning red, Pd+ orange-yellow. **17. L. radiosa** var. **subcircinata**
- Thallus K+ yellow, Pd−. **16. L. pruinosa**
3. Apothecia rare and few. Spores 10–14 × 5–7 μ. **14. L. bolcana**
- Apothecia numerous and crowded. Spores smaller than above. **15. L. muralis**

14. Lecanora bolcana (Poll.) Poelt, Mitt. Bot. Staatssamml. München, 19–20: 505 (1958). *Lecidea bolcana* Poll., Giorn. Fis. Chim. Stor. Nat. Pavia 9: 178 (1816). [Plate XI: 2]

Thallus more or less orbicular with a distinct lobate periphery when single, or several thalli fused to lose their original shape and form a continuous crust with only the lobes of the outermost colonies conspicuous; lobes up to 3 mm broad, as long as they are broad or shorter, usually plane, incised irregularly and in various directions, sometimes the lobes of two adjacent colonies overlapping; centre areolate; areoles plane, contiguous, irregular in shape and size, 0.5–1.5 mm; the bigger areoles with 1 or more blind incisions; adventive lobules usual in the centre and periphery; areoles, lobes and incisions surrounded by a thick, black rim; surface yellowish green, epruinose, dull. Cortex of variable thickness, densely cellular, upper part yellowish and covered by a thin, transparent, amorphous stratum. Medulla greyish-nubilated.

Apothecia rather sparse, at first immersed, later adnate, single or in groups and pressed against each other, roundish or shapeless. Disc usually plane, brown, surrounded by a slightly prominent more or less crenulate thalline margin; interior pale, concolourous with the thallus, exterior black; in section, exterior layer composed of thick, blackish, short-celled hyphae; median layer a greyish, nubilated continuation of the medulla. Exciple colourless, paraphysoid. Hymenium colourless, 60–70 μ thick, including the brownish epithecium. Hypothecium of densely aggregated colourless hyphae, subtended by groups of algae. Paraphyses coherent. Spores 8, simple, colourless, ellipsoid, 10.2–14 × 5–7 μ.

Reactions: Usually negative, sometimes K± yellowish; contains traces of usnic acid; spores I+ green; asci and subhymenial region I+ blue.

Habitat: On basalt. Upper Galilee.

Distribution: Mediterranean, somewhat extending into adjacent localities of the Euro-Siberian and Irano-Turanian regions.

15. Lecanora muralis (Schreb.) Rabh., Deutschl. Kryptog.- Fl. 2: 52 (1845). *Lichen muralis* Schreb., Spic. Fl. Lips. 130 (1771). [Plate XI: 3]

Thallus at first rosette-shaped, 3–4 cm across, closely attached to the substrate, greenish yellow, epruinose and somewhat shiny or slightly pruinose along the margins, later expanding over 20 cm or more and densely covered with apothecia, except for the lobate periphery; lobes 2–4 mm long and 0.5–1.5 mm broad, repeatedly re-branched, plane or moderately convex; centre areolate; areoles of irregular size and shape, scattered and lobulate or simple and contiguous, often black-rimmed.

Apothecia 0.5–1.5 mm across, at first immersed, later sessile. Disc pale brown, brown or blackish brown at first, later black-rimmed, naked or slightly pruinose, usually plane. Exciple pale, sometimes disappearing later. Thalline margin pruinose, irregular, moderately prominent, later disappearing. Hymenium and hypothecium colourless, variable in size, with a distinct subhymenial zone in between. Hypothecium subtended by a thick algal layer overlying a dense crystalline zone. Epithecium yellowish brown. Paraphyses coherent, uniform. Spores 8, simple, colourless, ellipsoid, 5–10 × 2–5 μ.

Reactions: Thallus K–, C–, Pd–; contains traces of usnic acid; hymenium I+ blue.

Habitat: On basalt and quarzolithic dolomite. Upper and Lower Galilee, Mt. Gilboa.

Distribution: Mainly Mediterranean; also found in the Euro-Siberian region and in N. America.

16. Lecanora pruinosa Chaub., in Saint-Amans, Fl. Agenaise 495 (1821). [Plate XII: 3]

Thallus a rosette, 1–3 cm across, many colonies usually approaching but not fusing, in close contact with the substrate; lobes radially arranged, *c.* 2 mm long, *c.* 0.5 mm broad at the base and *c.* 1 mm at the ends, more or less uniform in size and shape, plane to somewhat convex, contiguous, not overlapping; centre somewhat plane-areolated, whitish, greyish or rarely pale yellowish green and heavily pruinose. Cortex obscure, *c.* 30 μ thick. Algal layer interrupted, 50–65 μ thick. Medulla opaque with crystals and substrate particles.

Apothecia numerous, covering almost the entire central area, usually deformed by mutual pressure, 0.5–1.5 mm across, sessile. Disc pale or dark brown, pruinose or naked, plane or convex. Thalline margin persistent, entire or irregularly crenulate, slightly prominent, concolourous with the thallus. Hymenium colourless, 50–60 μ thick. Epithecium brownish, 10–15 μ thick. Hypothecium colourless. Paraphyses articulate; tips clavate with yellowish caps. Spores 8, simple, colourless, ellipsoid, 6–10 × 3–4 μ.

Reactions: Thallus K+ yellow, C+ orange, Pd–; apothecia K–; contains three unknown substances; hymenium I+ blue.

Habitat: Common on calcareous rocks and stones. Upper Galilee, Mt. Carmel, Coast of Carmel, Shefela, Judean Mountains.

Distribution: Mainly Mediterranean.

17. Lecanora radiosa (Hoffm.) Schaer. var. **subcircinata** (Nyl.) Zahlbr., Ann. k. k. Naturh. Hofmus. Wien 4: 354 (1889). *Lecanora subcircinata* Nyl., Flora 56: 18 (1873). [Plate XII: 2]

Thallus rosette-shaped, easily detached from substrate, ash-grey, sometimes sordid or reddish brown at the lobate region, usually pruinose; lobes radiose-plicate, long and finger-shaped or short and broad, simple or branched, with roundish or angular ends, contiguous, not overlapping, plane or convex; centre areolate, finally covered by apothecia. Cortex 20–25 μ thick, covered by an amorphous layer of dead cells 5–8 μ thick, or by a layer of dark crystals *c.* 20 μ thick. Algae arranged in groups. Medulla white, in section nubilated by crystals.

Apothecia abundant, 1 or several to each areole, immersed, at first roundish, later made angular by pressure, 0.5–1 mm across. Disc concave to plane, black, epruinose. Thalline margin thin, entire, slightly prominent, concolourous with the thallus. Epithecium greenish brown. Hymenium colourless, gelatinous, separated from the colourless hypothecium by a distinct subhymenial zone. Paraphyses agglutinate, septate, simple, with 4–5 upper cells shorter and thicker than the rest. Spores 8, simple, colourless, ovoid or ellipsoid, with 1–2 oil drops, 10–15 × 5–6 μ.

Reactions: Thallus K+ yellow gradually turning red, Pd+ orange-yellow; medulla K+ yellow turning brown; contains norstictic acid* (Pl. XXVI: 3); hymenium I+ green-blue soon turning reddish brown; hypothecium I+ green-blue.

Habitat: Very common on calcareous rocks. Upper and Lower Galilee, Dan Valley, Judean Mountains.

3. OCHROLECHIA Mass.

Thallus crustose, uniform, areolate, firmly attached to the substrate. Cortex thin, plectenchymatous. Phycobiont *Cystococcus* (Verseghy, 1962). Apothecia lecanorine. Margin generally thick and prominent. Hymenium colourless. Hypothecium yellowish. Paraphyses septate, branched and intricate. Spores 2–8, large, simple, transparent, colourless, ellipsoid or ovoid.

1. Ochrolechia parella (L.) Mass., Ricerch. Auton. Lich. 32 (1852). *Lichen parellus* L., Mantissa 1: 132 (1767). [Plate XII: 4]

Thallus spreading, up to 20 cm broad, indeterminate, *c.* 1 mm thick, greyish or creamy white, deeply cracked-areolate to warty-granular.

Apothecia densely crowded, usually several, of which 1–2(–3) mature, innate on one areole, sessile, orbicular; mature apothecia 2–2.5 mm across, more or less level with the thallus; apothecia development more uniform in warty-granular areas. Disc plane with a thick white pruina. Thalline margin thick, entire, creamy white, somewhat prominent. Hymenium colourless, 160–210 μ thick. Epithecium dull yellow. Hypothecium not distinct. Subhymenium 40–60 μ thick, subtended by an algal layer continuous to the margin. Paraphyses slender, branched and loosely intricate. Spores 6–8, colourless, simple, ovoid, 48–53 × 25–28 μ.

Reactions: Thallus K–, C–, Pd–; apothecia (disc) K–, C+ red, Pd–; thallus contains variolaric acid, apothecia variolaric and lecanoric acid (Pl. XXV: 2).

Habitat: Abundant on basalt. Upper Galilee.

* Erroneously published as containing salazinic acid (Galun & Lavee, 1966).

Distribution: Mediterranean and Medio-European, expanding into adjacent localities in Atlantic and Boreal regions.

4. Solenopsora Mass.

Thallus crustose, periphery lobate-effigurate, distinctly corticate above, attached to the substrate by medullary hyphae. Apothecia superficial, lecanorine, usually with both exciple and thalline margin. Paraphyses simple. Spores 8, colourless, 1-septate with thin cell walls, ellipsoid or fusiform. Mediterranean.

1. Thallus olive-brown. Medulla Pd–. **3. S. montagnei** var. **calcarea**
– Thallus white or grey. Medulla Pd+ 2
2. Thallus white. Medulla Pd+ orange. **1. S. candicans**
– Thallus grey. Medulla Pd+ yellow turning red at certain points. **2. S. cesatii** var. **grisea**

1. Solenopsora candicans Stein., Oester. Bot. Z. 68: 303 (1919). [Plate XII: 6]

Thallus orbicular, up to 7 cm across, white, farinose, lobate-effigurate at the periphery; lobes closely appressed to the substrate, confluent; centre verrucose-subareolate. Cortex *c.* 30 μ thick, more or less uniform, upper part nubilated by greyish granular deposits, lower part hyphose, colourless. Algal layer continuous and more or less uniform. Medulla densely intricate and nubilated by greyish granular particles.

Apothecia many, usually scattered, up to 1.5 mm across, sessile. Disc plane, dark brown or black, naked or thinly pruinose. Exciple distinct, entire, blackish, uniform in width with the hypothecium; hyphae parallel, colourless except for the dark brown pigmented outermost cells. Thalline margin white, pruinose, at first entire, later reflexed and almost excluded. Hymenium 50–60 μ thick, colourless except for a thin fuliginous epithecium. Hypothecium 90–100 μ thick, colourless, densely aggregated, subtended by a thick, continuous algal layer. Paraphyses simple, capitate; apical cell with a thick brown-pigmented cell wall. Spores 8, colourless, 2-celled, ellipsoid-oblong, 12–14 × 3–4 μ.

Reactions: Thallus K–, C–; medulla Pd+ orange; contains pannarin and zeorin.

Habitat: On oolithic limestone. Judean Mountains.

Distribution: Mediterranean; also found in the dry, warm localities of the Medio-European region and in N. America.

2. Solenopsora cesatii Zahlbr. var. **grisea** Bagl., Comment. Soc. Crittog. Ital. 1 (3): 121 (1862).

Thallus orbicular, lobate-effigurate, several cm across, greyish white, heavily pruinose; lobes 1–2 mm long, 0.3–1 mm broad, irregularly divided, convex-plicate with rounded ends; central lobules isidia-like, bullate, often splitting open but not truly soraliate. Medulla white.

Apothecia fairly numerous or sometimes rare, up to 1.5 mm across, adnate. Disc plane, dark brown, epruinose. Thalline margin concolourous with the thallus, white-pruinose, at first thick and entire, later partly excluded and deformed. Exciple persistent, fan-shaped, 20–50 μ thick; interior colourless; exterior cells yellowish brown and with crystalline deposits. Hymenium 30–45 μ thick, colourless. Epithecium 5–15 μ thick, yellowish. Hypothecium compactly hyphose, colourless, 60–90 μ thick, subtended by a continuous algal layer which does not reach the height of the hymenium in mature apothecia but is reflexed into the almost excluded margin. Paraphyses simple, gradually thicker towards the apex. Spores 8, colourless, 2-celled; ellipsoid and fusiform spores in one and the same apothecium, 8.5–12 × 3–5 μ.

Reactions: Thallus and medulla K–, C–, KC–; medulla Pd+ orange, turning red in certain spots; contains an unknown substance (Pl. XXIV: 2); hymenium I+ blue.

Habitat: On oolithic limestone, spreading onto mossy soil. Mt. Carmel, Shefela, Judean Mountains.

Distribution: Mediterranean.

3. Solenopsora montagnei Choisy et Werner var. **calcarea** Schaer., Enum. Crit. Lich. Europ. 63 (1850).

Thallus irregularly spreading, olive-brown, somewhat pruinose, especially along the periphery, minutely squamulose-areolate; squamules either flat, roundish, laciniate and overlapping, with occasionally raised margins, or wart-like, bullate; peripheral lobes 1–2 mm long, 2–3 mm broad, more or less plane, roundish, irregularly laciniate. Cortex more or less uniform, colourless except for an external brownish region, covered by a thick amorphous layer. Algal layer continuous and more or less uniform. Medulla nubilated by crystals and substrate particles.

Apothecia numerous, scattered or in groups, 1–1.5 mm across, constricted at the base. Disc brown, often with a very fine pruina, plane. Thalline margin thin, concolourous with the thallus, usually pruinose, finally partly excluded. Exciple well developed, persistent, colourless except for the exterior cells. Hymenium 50–60 μ thick, colourless. Epithecium gelatinous-fuliginous. Hypothecium compact, colourless, subtended by dispersed groups of algae. Paraphyses simple, capitate; apical cell brown. Spores 8, colourless, fusiform, mostly 1-septate, sometimes single-celled, 10–12 × 3–4 μ.

Reactions: Thallus and medulla K–, C–, Pd–; hymenium I+ blue.

Habitat: On oolithic limestone. Mt. Carmel.

Distribution: Var. *calcarea* has been reported only from S. France (Montpellier); the species is Mediterranean.

5. SQUAMARINA Poelt

Thallus rosulate or squamulose; lobes and squamules usually thick, broad and with somewhat rotund outlines, whitish or in various shades of yellow, brown and green, often entirely or partly pruinose. Upper cortex thick and uniform, of conglutinate, thick, perpendicular hyphae interspersed with yellowish granules. Algal cells arranged in a uniform, continuous layer. Apothecia lecanorine or sometimes biatorine. Epithecium granulose. Spores 8, simple, colourless, mostly ellipsoid. Substrate mainly calcareous.

1. Thallus rosulate 2
- Thallus squamulose 3
2. Pale brown, heavily pruinose. Medulla Pd–. **1. S. lentigera**
- Yellowish green, naked to moderately pruinose. Medulla Pd+ yellow (psoromic acid). **2. S. stella-petraea**
3. Squamules dispersed or adjacent, occasionally imbricate. Disc ochre. **3. S. gypsacea**
- Squamules always imbricate to overlapping. Disc brown to reddish brown. **4. S. crassa**

1. Squamarina lentigera (Web.) Poelt, Mitt. Bot. Staatssamml. München 19–20: 536 (1958). *Lichen lentigerus* Web., Spic. Fl. Goetting. 192 (1778). [Plate XIII: 1]

Thallus up to 4 cm across, 230–400 μ thick, always rosulate; peripheral lobes 2–3 mm long, 0.5–2 mm wide, usually contiguous, obtuse; central part subareolate or of partly imbricate squamules with white, moderately raised and free margins; upper side pale brown, with a white pruina *c.* 30 μ thick; lower side pale, adpressed to the substrate by medullary hyphae 2.5–3 μ thick.

Apothecia rather rare, centrally located, sessile, up to 2 mm across. Disc yellowish brown, slightly pruinose, plane. Thalline margin concolourous with the thallus, heavily pruinose, 100–125 μ thick, prominent or level with the disc. Hymenium colourless, *c.* 100 μ thick, with a yellowish, granulose epithecium 15–20 μ thick. Hypothecium colourless, 130–150 μ thick, subtended by a continuous algal layer that usually does not reach the height of the hymenium. Spores 8, colourless, simple, ellipsoid, 15 × 6–7.5 μ.

Reactions: Thallus K± yellow, KC+ yellow, medulla Pd–, C–; contains usnic acid (Pl. XXV: 4).

Habitat: Common on loess. Judean Desert, C. and W. Negev.

Distribution: Common in steppe and desert areas of the Mediterranean and Irano-Turanian regions, extending into Euro-Siberian territories.

2. Squamarina stella-petraea Poelt, Mitt. Bot. Staatssamml. München 19–20: 540 (1958). [Plate XIII: 3]

Thallus 2–5 cm across, always rosulate; peripheral lobes up to 0.8 mm long and 1–2 mm wide, adjacent, occasionally imbricate, plane to slightly convex, with somewhat inflated tips; central part more or less even and areolate; upper side yellowish green, smooth or with a thin pruina; lower side pale, rather firmly attached to the substrate by thick medullary hyphae.

Apothecia numerous, about 2 mm across. Disc plane or concave, brown. Thalline margin concolourous with the thallus and slightly pruinose, thick and prominent or revolute. Hymenium and hypothecium as in *S. lentigera.* Spores 8–12 × 5–6 μ.

Reactions: Medulla Pd+ yellow; contains psoromic acid (Pl. XXV: 6).

Habitat: On calcareous rocks and on mossy soil. Upper Galilee.

Distribution: Mediterranean, extending into some Irano-Anatolian and Medio-European regions.

3. Squamarina gypsacea (Sm.) Poelt, Mitt. Bot. Staatssamml. München 19–20 : 539 (1958). *Lichen gypsaceus* Sm., Trans. Linn. Soc. London 1 : 81 (1791). [Plate XIII : 2]

Thallus indeterminate-squamulose; squamules dispersed, contiguous or partly imbricate, 2–8 mm across, thick, roundish, somewhat plane or lacunose, olive-green with a somewhat raised white margin; lower side black, cracked, with black pseudorhizines.

Apothecia few, one–several (or none) per squamule, broadly adnate, up to 4 mm across. Disc pale yellowish or testaceous, plane or slightly concave. Exciple thin and persistent. Thalline margin thin and evanescent. Epithecium granulose, yellowish. Hymenium *c.* 100 μ thick, colourless, subtended by a distinct subhymenium 80–125 μ thick. Hypothecium 200–250 μ thick, colourless. Spores 8, simple, colourless, ellipsoid, 10–12 × 5–5.5 μ; mature spores rare.

Reactions: Thallus and medulla K–; medulla Pd+ yellow; contains psoromic acid (Pl. XXV : 6).

Habitat: On calcareous rocks. Upper Galilee.

Distribution: Mediterranean, Irano-Turanian, Medio-European.

4. Squamarina crassa (Huds.) Poelt, Mitt. Bot. Staatssamml. München 19–20 : 544 (1958). *Lichen crassus* Huds., Fl. Angl. ed. 2, 2 : 530 (1778).

Thallus very variable in shape and size, squamulose, occasionally rosulate when young; squamules 1–5 mm across, irregular, 300–650 μ thick, more or less overlapping or imbricate, loosely or closely attached to the substrate, plane or discoid or plicate; margins usually white, entire or crenate-lobate, raised or deflexed; upper side yellowish or greenish brown, or brown or entirely white-pruinose; lower side pale or dark brown, sometimes with pseudo-rhizines.

Apothecia sometimes rare to numerous, laminal, slightly constricted at the base, 0.5–1.5 mm across. Disc pale or dark or reddish brown, naked or pruinose, at first concave, later plane or convex. Thalline margin 80–145 μ thick, persistent or disappearing. Hymenium colourless, 130–160 μ thick. Epithecium yellow-brown, *c.* 15 μ thick. Hypothecium colourless, 75–190 μ thick. Spores 8, simple, colourless, ellipsoid, 7.5–13.5 × 3–4.5 μ.

Var. **crassa** f. **crassa** Poelt [Plate XIII : 4]. Medulla Pd+ deep yellow (psoromic acid).

Habitat: On loess or gypsous soil spreading over mosses and calcareous rocks. Upper, Lower and Coastal Galilee, Dan Valley, Mt. Carmel, Coast of Carmel, Shefela, Judean Mountains, Judean Desert, W. Negev.

Distribution: Common in dry and warm localities of the Mediterranean, Irano-Turanian and Medio-European regions.

F. **pseudocrassa** (Matt.) Poelt. *Lecanora lentigera* var. *pseudocrassa* Matt., Ber. Deutsch. Bot. Ges. 58 : 352 (1940). Medulla Pd–.

Habitat: On calcareous rocks. Upper and Lower Galilee, Mt. Carmel, Sharon Plain, Judean Mountains, C. Negev.

F. **iberica** (Matt.) Poelt. *Lecanora crassa* f. *iberica* Matt., Ber. Deutsch. Bot. Ges. 58 : 348 (1940). Medulla Pd+ red.

Habitat : On mossy soil. Upper Galilee.

6. Candelariella Müll. Arg.

Thallus crustose, warty-granular or areolate, uniform or lobate-effigurate, pale to deep yellow, in close contact with the substrate. Cortex paraplectenchymatous, containing stictaurin (K–). Phycobiont *Pleurococcus* (Hakulinen, 1954) (Fig. 1 C). Apothecia sessile, lecanorine. Epithecium granulose. Spores 8–32, colourless, simple or rarely 2-celled, thin-walled.

1.	Asci with 16–32 spores.	**2. C. vitellina**
–	Asci with 8 spores	2
2.	Thallus granular, lobate at the periphery.	**1. C. medians**
–	Thallus areolate, not lobate.	**3. C. minuta**

1. Candelariella medians (Nyl.) A. L. Sm., Monogr. Brit. Lich. 1 : 228 (1918). *Placodium medians* Nyl., Bull. Soc. Bot. France 9 : 262 (1862). [Plate XII : 5]

Thallus 3–5 cm across; centre becoming a warty-granular, dark grey, shapeless crust; periphery lobate; lobes yellow, 1–2 mm long, 0.2–1 mm broad, occasionally lobulate, adnate, more or less coherent. Cortex 15–22 μ thick, covered with an amorphous layer 5–8 μ thick.

Apothecia sparse, 0.3–0.5 mm across. Disc plane, at first yellow, later greyish. Thalline margin entire, slightly prominent, concolourous with the lobate region. Hymenium and hypothecium colourless, 60–80 μ thick. Epithecium granulose, at first yellow, later obscure. Spores 8, colourless, ellipsoid or bean-like, with one or several oil drops, usually simple, but some appearing 2-celled, 8.5–14 × 3.5–4 μ.

Reactions : Thallus K–, C–, Pd–; contains stictaurin (Pl. XXVIII : 1); hymenium I+ blue.

Habitat : On soft calcareous stones. Upper Galilee.

Distribution : Mediterranean and Medio-European.

2. Candelariella vitellina (Ehrh.) Müll. Arg., Bull. Herb. Boiss. 2 (App. 1) : 47 (1894). *Lichen vitellinus* Ehrh., Plant. Crypt. Exs. 155 (1785).

Thallus fragmentary, areolate; areoles dispersed, closely adnate, warty, yolk-yellow, 0.2–0.4 mm.

Apothecia numerous but not aggregated, up to 0.5 mm across. Disc plane, greenish yellow; margin thick, entire, prominent, concolourous with the thallus. Epithecium granular, yellowish. Hypothecium colourless, radiating into a fan-shaped exciple. Paraphyses aseptate, colourless, straight and free. Spores 16–32, mostly 20–30, ellipsoid or slightly curved, with 1–2 oil drops, 11–12 × 5–6 μ.

Reactions : Thallus K–, C–, Pd–; contains stictaurin (Pl. XXVIII : 1); hymenium I+ blue; hypothecium and exciple I–.

Habitat : On basalt. Upper and Lower Galilee, Upper Jordan Valley, Judean

Mountains. Usually not in direct contact with the basalt, but growing on fragments of blue-green alga lichens and on free blue-green algae.

Distribution: Cosmopolitan.

3. **Candelariella minuta** Reichert et Galun, Bull. Res. Counc. Israel 9 D: 135 (1960).

Thallus in small roundish patches, 1.5–2 cm, or intermingled with other lichens, areolate; areoles greyish, minute, with a plane base and prominent margin, arranged in more or less parallel rows.

Apothecia numerous, up to 0.5 mm across. Disc plane or concave, lemon-yellow; thalline margin prominent, pale, at first entire, later incised. Exciple 8–15 μ thick, distinctly hyphose. Hymenium colourless, 70–90 μ thick. Hypothecium colourless, 75–95 μ thick. Epithecium 6–8 μ thick, brown, granular. Spores 8, colourless, ellipsoid or bean-like, with both ends rounded or one end pointed, 12–18 × 4.5–6 μ.

Reactions: Thallus K–, C–, Pd–; contains stictaurin (Pl. XXVIII: 1); hymenium I+ blue.

Habitat: On dolomite and limestone. W. Negev.

PARMELIACEAE

Thallus foliose to subfruticose, stratified, corticate above and below, attached to the substrate by rhizines, prothallus or by wrinkles of the lower side. Phycobiont belonging to the Chlorophyceae. Apothecia laminal, marginal or terminal, sessile or shortly stalked, discoid, lecanorine. Spores usually 8, simple, colourless.

PARMELIA Ach.

Thallus foliose, laterally expanded, loosely to firmly attached to the substrate by rhizines. Upper cortex and lower cortex paraplectenchymatous. Phycobiont *Trebouxia* (Fig. 1 A). Medulla plectenchymatous, white. Apothecia laminal, lecanorine. Hypothecium colourless. Spores 8, simple, colourless.

1. Thallus brown — 2
– Thallus not brown — 4
2. Corticolous. Medulla C+ red (lecanoric acid). — **1. P. glabra**
– Saxicolous. Medulla C– — 3
3. Thallus isidiate, KC+ red (glomelliferic acid). — **2. P. glomellifera**
– Thallus not isidiate, KC–. — **3. P. perrugata**
4. Thallus grey. Medulla C+ red (lecanoric acid). — **4. P. tiliacea**
– Thallus yellowish green. Medulla K+ yellow→red, Pd+ orange (salazinic acid). — **5. P. tinctina**

1. **Parmelia glabra** Nyl., Flora 55: 548 (1872). [Plate XIV: 1]

Thallus foliose, rosulate, densely rhizinate; rosettes up to 8 cm across, closely adherent to the substrate; upper surface at first olive-brown and glistening, finally dark brown and dull with many very small transparent simple hairs; central lobes irregularly overlapping, plicate or wrinkled and incised to form isidia-like lobules; peripheral lobes roundish with entire white, somewhat ascending margins; lower surface black and somewhat paler at the marginal zone; rhizines dark brown.

Apothecia numerous, laminal, up to 5 mm across. Disc cup-shaped, brown, glistening in young apothecia, epruinose. Thalline margin rather thick, concolourous with the thallus, denticulate, with the white medulla breaking through. Hymenium colourless, 60–70 μ thick. Epithecium gelatinous, yellowish. Hypothecium pale yellow, subtended by a continuous algal layer. Spores 8, simple, colourless, ellipsoid, 15–18 × 6–8 μ.

Reactions: Medulla C+ red; contains lecanoric acid (Pl. XXV: 2).

Habitat: On bark of old olive trees. Upper Galilee.

Distribution: Medio-European; also found in the W. part of N. America; reported from India.

2. Parmelia glomellifera Nyl., Flora 64: 453 (1881). [Plate XIV: 4]

Thallus foliose, isidiate, spreading over several cm, dark olive-brown to dark brown, shiny, minutely wrinkled and bulging, somewhat cracked, adnate; lobes long and narrow, deeply incised, flat or partly folded, contiguous or overlapping or irregularly entangled, with a narrow darker zone along the margins and dentate at the edges; lower side blackish, brown toward the periphery, rhizinate; rhizines blackish, short. Isidia concolourous with the thallus or somewhat darker, laminal, occupying almost the entire centre of the thallus, arranged in more or less uniform groups, simple or coralloid, some bursting at the tip and becoming white-sorediose.

Sterile.

Reactions: Medulla C–, Pd–, KC+ red; cortex K+ at first greenish blue, turning reddish brown; contains glomelliferic acid (Pl. XXVIII: 2).

Habitat: On basalt. Upper Galilee.

Distribution: Euro-Siberian and Mediterranean.

3. Parmelia perrugata Nyl., Flora 68: 295 (1885). [Plate XIV: 3]

Thallus foliose, rosulate or irregularly spreading, 6–10 cm across, adnate; lobes flattened and rigid, contiguous or somewhat overlapping, irregularly incised; centre rough and wrinkled because of peculiar densely aggregated warts or papillae, each with a depression at the apex giving them the appearance of premature apothecia; upper surface dark brown, dull, bluish-pruinose along the margin of the peripheral lobes; lower side black, cracked, brownish and shiny towards the periphery, rhizinate; rhizines blackish, some white-penicillate.

Apothecia many, centrally located, 2–8 mm across, sessile. Disc dark brown, at first plane, later wrinkled and folded. Thalline margin thin and entire, dark brown. Hymenium and hypothecium colourless. Epithecium gelatinous, yellowish. Spores 8, simple, colourless, ovoid, 6.5–8 × 2.5–4 μ.

Reactions: Medulla K–, C–, KC–, Pd–.

Habitat: On basalt. Upper and Lower Galilee, Upper Jordan Valley, Mt. Gilboa, Judean Mountains, Shefela.

Distribution: N., E. and W. Mediterranean.

4. **Parmelia tiliacea** (Hoffm.) Ach., Method. Lich. 215 (1803). *Lichen tiliaceus* Hoffm., Enum. Lich. 96 (1784). [Plate XIV: 2]

Thallus foliose, in rosette form or spreading, closely attached in the centre; lobes loose, 0.5–1.5 cm long, 0.2–1 cm broad, rotund, wavy, contiguous or partly imbricate; margins slightly raised or bent down or flat; upper surface ash-grey to silver-grey, shiny or slightly pruinose, isidiate; isidia densely concentrated in the centre, cylindrical, simple or rarely branched, brown or with brown tips; lower side dark brown to black, getting paler and shiny towards the periphery, rhizinate; rhizines long, black, simple, fewer and shorter on the peripheral lobes.

Apothecia sometimes numerous, sometimes sparse, up to 2 mm across. Disc pale brownish yellow, shiny, concave. Thalline margin thick, concolourous with the thallus, entire or incised in old apothecia, sometimes isidiate. Hymenium colourless, subtended by a distinct subhymenium. Epithecium gelatinous, yellowish brown. Hypothecium colourless. Spores 8, colourless, simple, ellipsoid, usually with 2 vacuoles, 8–12 × 5–8 μ.

Reactions: Thallus K+ yellowish, C–; medulla K–, C+ red; contains lecanoric acid (Pl. XXV: 2) and atranorin (Pl. XXV: 1).

Habitat: Common on *Quercus calliprinos,* olive trees and basalt. Upper Galilee.

Distribution: Euro-Siberian and Mediterranean.

5. **Parmelia tinctina** Mah. et Gill., Bull. Soc. Bot. France 72: 860 (1925).

Thallus foliose, rosulate, up to 10 cm across, adnate; lobes 1–2 cm long and 1–2 mm wide, broadening up to 5–6 mm at the tips, irregularly branched, contiguous or overlapping, in some places discrete, somewhat wavy and wrinkled towards the centre; upper surface greenish yellow, blackening in the centre, isidiate; isidia mainly in the centre, concolourous with the thallus, some with darker apices, simple, subglobose; lower side black or dark brown, densely rhizinate; rhizines black, simple.

Apothecia sparse, laminal, sessile, up to 5 mm across. Disc at first deeply concave, later flattened, golden brown, epruinose. Thalline margin thin, at first entire, finally crenate or wrinkled, sometimes isidiate. Epithecium brown, granular, 8–12 μ thick. Hypothecium colourless, 120–160 μ thick, subtended by a more or less continuous algal layer. Hymenium colourless, 80–100 μ thick. Spores 8, colourless, simple, ellipsoid, 12–14 × 7–9 μ.

Reactions: Medulla K+ yellow turning red, Pd+ orange; contains usnic acid (Pl. XXV: 4), atranorin (Pl. XXV: 1) and salazinic acid (Pl. XXVI: 2).

Habitat: On flint. Rare. Upper Galilee.

Distribution: W. Scandinavia, England, France, S. Europe and N. Africa; also known from India, E. Asia, S. Africa and N. America (Hale, 1964).

USNEACEAE

Thallus fruticose, tufted or pendulous, attached to the substrate by a holdfast; branches radial or dorsiventral, corticated on all sides. Phycobiont belonging to the Chlorophyceae. Apothecia discoid, lecanorine. Spores 1–8, colourless or rarely brown, simple or septate.

1. Structure dorsiventral; upper and lower surfaces differing in colour. Spores simple. **1. Evernia**
– Structure radial; colour alike on both sides. Spores 1-septate. **2. Ramalina**

1. EVERNIA Ach.

Thallus fruticose, erect or partly pendulous or decumbent, attached by a holdfast, branched, differently coloured above and below, dorsiventral. Cortical hyphae perpendicular to the surface. Algae mostly under the upper cortex, except at the apices. Apothecia lecanorine, lateral or terminal. Spores 8, simple.

Evernia prunastri (L.) Ach., Lichgr. Univ. 442 (1810). *Lichen prunastri* L., Spec. Plant. 2 : 1147 (1753). [Plate XV : 1]

Thallus fruticose, partly erect or decumbent, rather soft, repeatedly branched, up to 5 cm high; fronds 1–4 mm wide, dorsiventral, flat or somewhat furrowed with margins softly bent down; apices almost cylindrical with pointed tips; upper surface pale green, sorediate; lower side white; soredia plentiful, marginal. Cortex uneven; cortical hyphae perpendicular to the surface. Algal layer beneath the upper cortex only, except for an algal ring around the medulla at the young cylindrical apices. Medulla white, loose; medullary hyphae 4–5 μ thick.

Sterile.

Reactions: Thallus K± slightly yellow; thallus and medulla C–, Pd–; contains atranorin (Pl. XXV : 1), usnic acid (Pl. XXV : 4) and evernic acid (Pl. XXIV : 6).

Habitat: On branches of *Rhamnus* and *Quercus*. Upper Galilee.

Distribution: Common in temperate regions.

2. RAMALINA Ach.

Thallus fruticose, tufted, erect or pendulose, attached by a basal holdfast; branches alike on both sides, radial, compressed or terete. Cortex often strengthened by an internal layer of longitudinal hyphae, otherwise all cortical hyphae perpendicular to the surface. Phycobiont *Trebouxia* (Fig. 1 A). Apothecia marginal or terminal, lecanorine. Spores 8, colourless, 1-septate. Usnic acid present in most species, often together with other substances.

1. Thallus sorediate, usually sterile or rarely with apothecia 2
– Thallus without soredia but with apothecia. **1. R. fastigiata**
2. Cortex single-layered; all cortical hyphae perpendicular to the axis 3

- Cortex double-layered; exterior cortical hyphae perpendicular, interior hyphae parallel to the axis 4
3. Thallus saxicolous, rigid. Apothecia usually absent. Medulla K+ yellow → red, Pd+ dark yellow; contains norstictic acid. **2. R. maciformis**
- Thallus corticolous, softer. Apothecia rare. Medulla K–, Pd–. **3. R. duriaei**
4. Soredia mainly terminal, partly laminal and marginal. Medulla and soredia K–, Pd–; contains evernic acid. **6. R. pollinaria**
- Soredia marginal 5
5. Medulla and soredia K–, Pd–, or K+ yellow → red, Pd+ yellow → red; contains protocetraric acid. **4. R. farinacea**
- Medulla and soredia K+ yellow → red, Pd+ yellow → orange; contains norstictic and salazinic acid. **5. R. reagens**

1. Ramalina fastigiata (Pers.) Ach., Lichgr. Univ. 603 (1810). *Lichen fastigiatus* Pers., Neue Ann. d. Bot. 1 : 156 (1794). [Plate XV : 3]

Thallus consisting of erect, more or less rigid bushes, 1–3 cm high, 2–4 cm wide, greyish green, turning yellowish brown with time; branches 0.5–2 cm long and 1–4 mm wide, flat and longitudinally lacerated, usually with many small, thin and ramose branchlets with pointed forked ends; branches becoming inflated near the terminal apothecia; soredia and isidia absent. Cortex colourless, double-layered; exterior layer thin, composed of hyphae perpendicular to the axis, interior layer thicker, composed of hyphae parallel to the axis. Algae only on one side of the medulla, except at the ends of the fronds where an algal ring encloses the white medulla.

Apothecia terminal, 1–5 mm across. Disc plane, pale yellowish brown, slightly pruinose. Thalline margin thin, level with the disc, concolourous with the thallus. Hymenium 40–50 μ thick, colourless. Epithecium sordid, granular. Hypothecium colourless. Spores 8, colourless, 1-septate, often curved, 9–17 × 4.5–7 μ.

Reactions: Thallus and medulla K–, KC–, Pd–; contains usnic acid (Pl. XXV : 4).

Habitat: On olive trees. Rare. Upper Galilee.

Distribution: Euro-Siberian and Mediterranean; also found in N. America.

2. Ramalina maciformis (Del.) Bory, Dictionn. Class. Hist. Nat. 14 : 458 (1828). *Parmelia maciformis* Del., Descript. de l'Egypte 2 : 288 (1813). [Plate XV : 2]

Thallus consisting of erect, rigid tufts, 0.5–3 cm high and 1–8 cm wide, pale greyish green when fresh but changing to brownish with a faint reddish tinge with time, dull; tufts composed of compressed, broadly flattened fronds of very variable shape and size (1–25 mm in width and 0.5–3 cm in height), irregularly branched and subdivided, straight, undulate or twisted, usually wrinkled-reticulate on one side or rather often on both sides. Cortex often cracked along the reticulate network leaving the medulla exposed. Soredia numerous, urceolate, usually on one surface, but often on both, sometimes with secondary development of small laciniae from the soredia (Fig. 12). Cortex 65–70 μ thick, single-layered, composed of thin, colourless hyphae more or less perpendicular to the surface and embedded in a dense gelatinous stratum. Algal layer *c.* 50 μ thick, more or less uniformly enclosing the medulla. Medulla white, 120–140 μ thick.

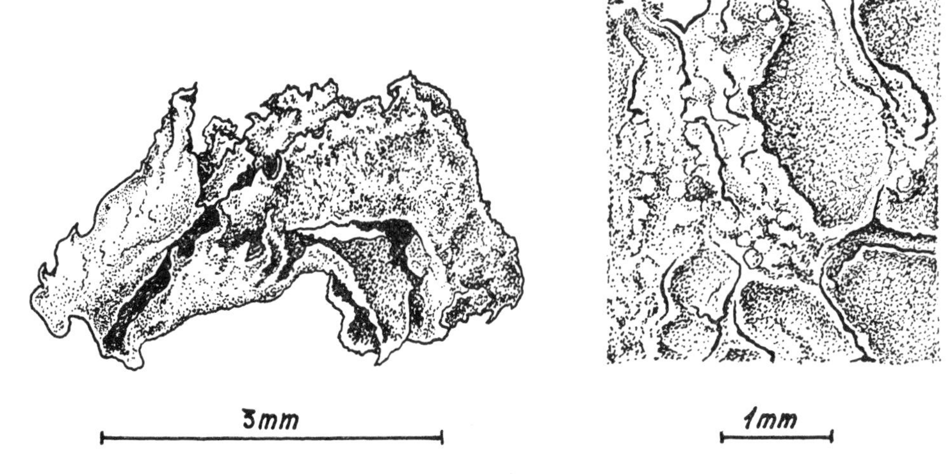

Fig. 12
Ramalina maciformis

Sterile.

Reactions: Medulla K+ yellow turning red, C–, Pd+ dark yellow; contains norstictic acid (Pl. XXVI: 3).

Habitat: Abundant on flint hammadas in C. and W. Negev, also on limestone and dolomite in Upper Galilee, Esdraelon Valley and Mt. Gilboa.

Distribution: Mediterranean, Irano-Turanian and Saharo-Arabian.

3. **Ramalina duriaei** (De Not.) Jatta, Monogr. Lich. Ital. 83 (1889). *R. pollinaria* var. *duriaei* De Not., Giorn. Bot. Ital. 1 (1): 216 (1846).

Thallus as in *R. maciformis* but softer and becoming pale yellowish brown with time.

Apothecia rare, up to 6 mm across, marginal or laminal, very shortly pedicellate. Disc deeply concave, pale greyish green, smooth. Thalline margin of the same colour, entire. Hymenium colourless, 40–75 μ thick. Hypothecium densely compact, colourless, 30–40 μ thick. Spores 8, 2-celled, colourless, ellipsoid, straight or somewhat curved, 10–15 × 3–4 μ.

Reactions: Thallus and medulla K–, C–, KC–, Pd–; hymenium I+ blue.

Habitat: Common on olive trees, oaks and *Rhamnus*. Upper and Lower Galilee, Mt. Carmel, Coast of Carmel, Shefela, Judean Mountains.

Distribution: Mediterranean, Medio-European and Atlantic.

4. **Ramalina farinacea** (L.) Ach., Lichgr. Univ. 606 (1810). *Lichen farinaceus* L., Spec. Plant. 1146 (1753). [Plate XV: 4]

Thallus consisting of erect or pendulous tufts of branches; branches sorediate, rather stiff, narrow (1–3 mm wide), more or less linear, attenuate, 3–5 cm high, compressed and channelled or almost cylindrical in younger parts, repeatedly dichotomously branched, pale green and glossy. Soredia mostly marginal, in roundish or ellipsoid patches. Cortex double-layered; exterior cortex composed of thin hyphae perpendicular to the axis, interior cortex composed of thicker longitudinal hyphae. Medulla white, loose. Algal cells mainly dispersed in the medulla.

Sterile.

Reactions: Medulla and soredia K–, Pd–, or K+ yellow turning red, Pd+ yellow turning red; contains protocetraric acid (Pl. XXIV: 1) and usnic acid (Pl. XXV: 4).

Habitat: On *Rhamnus palaestina* and *Quercus calliprinos*. Upper Galilee.

Distribution: Cosmopolitan.

5. **Ramalina reagens** * (B. de Lesd.) Culb., Rev. Bryol. et Lichenol. 34: 847 (1966). *R. farinacea* var. *reagens* B. de Lesd., Bull. Soc. Bot. France 67: 217 (1920).

Thallus as in *R. farinacea*.

Sterile.

Reactions: Medulla and soredia K+ yellow turning red, Pd+ yellow turning orange; contains salazinic acid (Pl. XXVI: 2), norstictic acid (Pl. XXVI: 3) and usnic acid (Pl. XXV: 4).

Habitat: On basalt. Upper Galilee.

6. **Ramalina pollinaria** (Ach.) Ach., Lichgr. Univ. 608 (1810). *Lichen pollinarius* Ach., Kgl. Vetensk. Akad. Nya Handl. 263 (1797). [Plate XV: 5]

Thallus consisting of dense, erect, small cushions, 1–3 cm in height; branches irregularly divided, 1–3 mm wide, more or less compressed, wrinkled, pale green, somewhat glossy, ending in delicately divided branchlets or in tips which burst into lumps of soredia. Soredia also distributed, in patches of irregular size and shape, almost down to the base of the branches. Cortex double-layered; exterior cortical hyphae perpendicular to the axis, embedded in a gelatinous stratum and inspersed below with fine yellowish granules; interior cortical hyphae longitudinal. Algae restricted to a continuous layer. Medulla white, loosely intricate.

Sterile.

Reactions: Thallus K–, C–; medulla Pd–; contains usnic acid (Pl. XXV: 4) and evernic acid (Pl. XXIV: 6).

Habitat: On basalt. Upper Galilee.

Distribution: Pluriregional.

PHYSCIACEAE

Thallus crustose, foliose or fruticose, stratified, corticate above or on both sides or ecorticate. Phycobiont belonging to the Chlorophyceae (*Trebouxia* in most cases). Apothecia lecanorine or lecideine. Spores brown, 2–4 celled, polarilocular or placodiomorph.

1. Thallus crustose 2
– Thallus foliose or fruticose 3

* Published as *R. subfarinacea* (Galun & Lavee, 1966).

2. Apothecia lecanorine. **1. Rinodina**
- Apothecia lecideine. **2. Buellia**
3. Upper cortex fibrous, composed of longitudinal hyphae more or less parallel to the surface 4
- Upper cortex cellular, para- or prosoplectenchymatous 5
4. Branches dorsiventral. **5. Anaptychia**
- Branches radial. **6. Tornabenia**
5. Thallus grey to brown. Upper cortex para- or prosoplectenchymatous, K–. Spore walls sculptured, very thick near the septum. **4. Physconia**
- Thallus whitish, pale grey or rarely brown. Upper cortex paraplectenchymatous, K+ yellow (rarely K–). Spore walls smooth, thickened at the apices and septum. **3. Physcia**

1. Rinodina (Ach.) S. Gray

Thallus crustose, uniform, stratified, ecorticate, attached to the substrate by medullary hyphae or by a hypothallus. Phycobiont *Trebouxia* (Fig. 1 A). Apothecia with both thalline margin and exciple. Hypothecium usually colourless or pale. Spores 8, brown, polarilocular or mischoblastiomorph (Fig. 13 A).

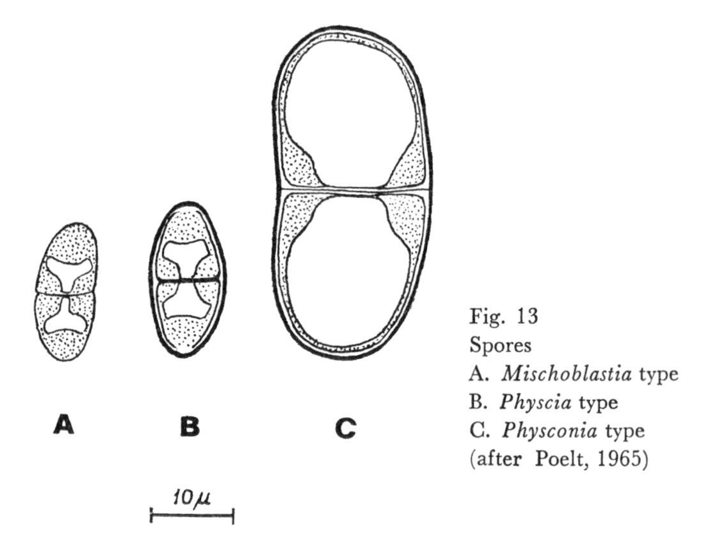

Fig. 13
Spores
A. *Mischoblastia* type
B. *Physcia* type
C. *Physconia* type
(after Poelt, 1965)

1. Saxicolous 2
- Corticolous 3
2. Apothecia subinnate, maximum diameter 0.5 mm. Thalline margin evanescent. Exciple paraplectenchymatous. **2. R. mediterranea**
- Apothecia superficial, up to 1.0 (–1.5) mm across. Thalline margin partially excluded. Exciple prosoplectenchymatous. **1. R. bischoffii** var. **aegyptiaca**
3. Thallus dark olive-green, almost black. Hypothallus absent. **3. R. carmeli**
- Thallus paler. Hypothallus greyish black. **4. R. magnussoniana**

1. Rinodina bischoffii (Hepp) Mass. var. **aegyptiaca** Müll. Arg., Rev. Mycol. 2 : 77 (1880). [Plate XVI : 1]

Thallus inconspicuous, composed of discrete, greyish, scurfy granules, mainly between the apothecia.

Apothecia numerous, sessile, 0.4–1.0(–1.5) mm across, dispersed or in groups. Disc black, naked or with a fine pruina, at first plane, later convex. Exciple black, thin, of colourless columnar hyphae (prosoplectenchymatous) with the exterior cells dark brown to black. Thalline margin greyish, entire, becoming partially excluded. Hymenium colourless, 90–100 μ thick. Epithecium dark brown. Hypothecium colourless, 60–80 μ thick; no algal cells beneath the central portion of the hypothecium. Paraphyses septate and furcate in the upper third only; apical cell enlarged and dark brown, sometimes 1–3 apical cells pigmented. Spores 8, brown, polarilocular, some slightly constricted at the septum, outer spore wall thin, thickened near the septum, 15–18 × 9–12 μ.

Reactions: Thallus K–, C–, Pd–; hymenium I+ blue → brown.

Habitat: On calcareous stones and flinty limestones. C. and W. Negev.

Distribution: Irano-Turanian and Saharo-Arabian.

2. Rinodina mediterranea Flagey, Cat. Lich. Algerie 40 (1896).

Thallus inconspicuous except for a few small, grey dispersed granules visible only under high magnification.

Apothecia numerous, scattered, minute, maximum diameter 0.5 mm, subinnate. Disc dark brown or black, naked or with a thin pruina, moderately convex. Thalline margin greyish, soon becoming excluded. Hymenium colourless, 65–90 μ thick. Epithecium brown. Hypothecium not distinctly separated from the hymenium, colourless, 60–100 μ thick. Exciple thin, dark brown, paraplectenchymatous; interior colourless; peripheral cells with dark brown cell walls, similar to and continuous with the capitate apical cells of the paraphyses. Spores 8, at first greenish then brown, polarilocular with rounded lumina, ellipsoid, spore walls thin but thickened near the septum, not constricted, 18–20 × 10.5–12 μ.

Habitat: On calcareous rocks. C. Negev.

Distribution: Known only from Algeria and Greece.

3. Rinodina carmeli Reichert et Galun, Bull. Res. Counc. Israel 6 D : 242 (1958).

Thallus thin to very thin, more or less determinate; periphery granular-verrucose; centre rimose, dark olive-green verging upon black. Hypothallus lacking.

Apothecia many, contiguous, adnate, 0.4–0.8 mm across. Disc blackish, pruinose, plane or somewhat convex. Thalline margin concolourous with thallus, prominent, crenulate, 75–150 μ thick. Exciple almost indistinct or very thin, laterally 6–10 μ, expanded to 30 μ above. Hymenium 100–125 μ thick. Epithecium fuliginous. Hypothecium 45–65 μ thick, colourless or faintly yellow. Paraphyses with broadened brownish apical cells. Spores 8, polarilocular, brown, lateral walls and septum thin (1–1.5 μ), apical walls thickened (3–3.5 μ), 15–18 × 7.5–9 μ.

Reactions: Thallus K+ yellowish; hymenium I+ blue; hypothecium I+ violet.
Habitat: On twigs of olive trees. Mt. Carmel.

4. Rinodina magnussoniana Reichert et Galun, Bull. Res. Counc. Israel 6 D: 242 (1958). [Plate XVI: 2]

Thallus very thin, minutely granular-verrucose or cracked, pale to somewhat dark greyish olive. Hypothallus greyish black.

Apothecia numerous to crowded, adnate, 0.3–0.7 mm across. Disc plane or sometimes convex, black and slightly pruinose. Thalline margin somewhat crenulate, concolourous with the thallus, *c.* 60 μ thick. Exciple more or less uniform, 7–16 μ thick, confluent with the hypothecium; external cells dark brown. Hymenium 90–110 μ thick, colourless. Epithecium dark brown. Hypothecium 75–90 μ thick, colourless or faintly yellow. Paraphyses with enlarged brown apices. Spores 8, mischoblastiomorph, greenish brown or brown, 15–18 × 7.5–9 μ.

Reactions: Thallus K+ yellow.
Habitat: On olive trees. Upper Galilee, Samaria.

2. Buellia De Not.

Thallus crustose, uniform or effigurate-lobate at the periphery; stratified, attached to the substrate by medullary hyphae or by a hypothallus. Cortex differentiated or ecorticate. Phycobiont *Trebouxia* (Fig. 1 A). Apothecia lecideine. Hypothecium dark. Spores 8, brown, 2–4-celled.

1.	Thallus effigurate-lobate. Cortex distinct	2
–	Thallus uniform. Cortex indistinct	3
2.	Soralia present. Medulla Pd+ yellow.	**6. B. canescens**
–	Soralia absent. Medulla Pd–, K+ yellow → red.	**5. B. zoharyi**
3.	Spores 1-septate	4
–	Spores 3-septate	5
4.	Thallus greyish green, soraliate.	**1. B. sorediosa**
–	Thallus white, without soralia.	**2. B. subalbula** var. **fuscocapitellata**
5.	Apothecia mainly innate. Hypothecium reddish brown. Medulla K–.	**3. B. epipolia**
–	Apothecia adnate. Hypothecium dark brown. Medulla K+ red in certain spots.	**4. B. venusta**

1. Buellia sorediosa Reichert et Galun, Bull. Res. Counc. Israel 9 D: 150 (1960).

Thallus crustose, very thin, more or less orbicular, greyish green with a yellow tinge, bordered by a black hypothallus 1–1.5 mm broad, areolate; areoles at first minute, contiguous, later bursting into punctiform soralia finally covering almost the entire thallus; centre sometimes splitting apart to reveal the black hypothallus.

Apothecia very rare, adnate, up to 0.7 mm across. Disc concave or plane, black, epruinose. Exciple black, prominent, 30–45 μ thick. Hypothecium 80–110 μ thick; upper part brown, lower part blackish brown. Hymenium 90–140 μ thick, colourless. Epithecium gelatinous, green. Paraphyses simple, uniform. Spores 8, brown, 2-celled, cells of unequal size, 16–18 × 6–7.5 μ.

Reactions: Thallus K+ yellow, thallus and medulla Pd−; hymenium I+ blue.
Habitat: On flint. Mt. Gilboa, N., W. and C. Negev.

2. Buellia subalbula (Nyl.) Müll. Arg. var. **fuscocapitellata** M. Lamb, Jour. Bot. 74: 350 (1936). [Plate XVI: 3]

Thallus crustose, effuse-indeterminate, white, irregularly rimose-areolate, areoles plane, 1–1.5 mm across, *c.* 0.5 mm thick. Hypothallus lacking.

Apothecia numerous, scattered, at first partly innate, later adnate, up to *c.* 0.5 mm across. Disc at first plane and marginate, later convex and immarginate, black, naked or often pruinose. Hymenium colourless, *c.* 80 μ thick. Epithecium dark brown. Hypothecium 100–160 μ thick, dark reddish brown. Paraphyses septate, branched above; apices clavate, fuco-capitate. Spores 8, dark brown, 2-celled, cell wall and septum thin and uniform, 9–15 × 5–6 μ.

Reactions: Thallus K−, C−, Pd−; medulla K+ red; hymenium I+ blue.
Habitat: On calcareous stones and on flint. C., W. and S. Negev.
Distribution: Irano-Turanian and Saharo-Arabian.

3. Buellia epipolia (Ach.) Mong., Bull. Acad. Intern. Geogr. Bot. 9: 242 (1900). *Lecidea epipolia* Ach., Lichgr. Univ. 186 (1810).

Thallus crustose, indeterminate, chalky white, rimose-areolate, 250–500 μ thick; delimited from other lichens by a black hypothalline line.

Apothecia numerous, crowded to dispersed, up to 0.8 mm across, mostly innate and level with the thallus, some emerging and with a thin pseudothalline margin. Disc plane to convex, black, pruinose. Exciple black, slightly raised or not. Hymenium colourless, 70–80 μ thick. Epithecium blackish brown. Hypothecium dark reddish brown. Paraphyses septate, occasionally furcate; 2 apical cells clavate, brown, capitate. Spores 8, at first greenish, later brown, 3-septate, 16–18 × 7–8 μ.

Reactions: Thallus and medulla K−, C−, Pd−; hymenium I+ blue.
Habitat: On calcareous stones. C. and S. Negev.
Distribution: S. Medio-European, Mediterranean, Irano-Turanian and Saharo-Arabian.

4. Buellia venusta (Körb.) Lett., Hedwigia 52: 244 (1912). *Diplotomma venustum* Körb., Parerg. Lich. 179 (1865). [Plate XVI: 4]

Thallus crustose, more or less determinate, whitish, rimose-areolate; areoles plane, farinose, 290–350 μ thick. Hypothallus lacking.

Apothecia numerous, up to 1 mm across, adnate. Disc black, often pruinose, at first plane and with a pseudothalline margin, later convex with only the black exciple visible. Hymenium colourless, 70–80 μ thick. Epithecium olive-brown. Hypothecium dark brown, 100–140 μ thick. Paraphyses septate, occasionally furcate; apical cells brown, capitate. Spores 8, 3-septate, at first greenish, later brown, straight or curved, 15–19.5 × 7.5 μ.

Reactions: Thallus K−, medulla K+ red in places; hymenium I+ blue.
Habitat: On calcareous stones. C. and S. Negev.

Distribution: Medio-European, Mediterranean, Irano-Turanian and Saharo-Arabian.

5. Buellia zoharyi Galun, sp. nov.

Thallus albidus ambitu effiguratus, lobis radiato-plicatis, contiguis; in parte centrali squamuloso-areolatus.

Apothecia sparsa, sessilia; disco nigro, primum plano et marginata demum convexo et immarginato; hymenium hyalinum; epithecium fuscum; hypothecium fusco-atrum; sporae 8-nae, fuscae, ellipsoideae, uniseptatae.

Thallus K± lutescens; medulla K+ lutescens demum rubescens, C–, Pd–. Atranorinum et acidum sticticum continens.

Thallus whitish, farinose, 345–450 μ thick, 1–2(–3) cm across, lobate-effigurate at the periphery; tips bluish grey; lobes radiate-plicate, contiguous, 3–5 mm long and 1–1.5 mm broad, firmly attached to the substrate; centre squamulose-areolate. Cortex consisting of an exterior amorphous greyish layer *c.* 45 μ thick, and a colourless paraplectenchymatous inner layer.

Apothecia few, sessile, up to 0.8 mm across. Disc black, at first plane and marginate, later convex and immarginate. Exciple 60 μ thick above, 35–40 μ laterally, externally black, internally colourless; peripheral cells blackish brown. Hymenium colourless, 60–95 μ thick. Epithecium dark brown. Hypothecium blackish brown, 90–110 μ thick. Paraphyses septate, simple, gradually becoming thicker towards the apex; tips capitate, brown. Spores 8, 1-septate, at first greyish green, later brown, ellipsoid, straight or occasionally curved, cell walls thin, 12–16.5 × 7.5 μ.

Reactions: Thallus K± yellow; medulla K+ yellow turning red, C–, Pd–; contains stictic acid (Pl. XXVI: 1) and atranorin (Pl. XXV: 1); hymenium I+ blue.

Habitat: On loess. W. and C. Negev.

6. Buellia canescens (Dicks.) De Not., Giorn. Bot. Ital. 1: 197 (1846). *Lichen canescens* Dicks., Fasc. Crypt. Brit. 1: 10 (1785). [Plate XVI: 5]

Thallus orbicular, rather loosely attached to the substrate, lobate-effigurate at the circumference; lobes radiate-plicate, contiguous, 1–4 mm long; centre areolate, soraliate, pale ash-grey, somewhat darker towards the periphery, pruinose, 250–500 μ thick at the centre, 150–200 μ thick at the lobate region. Cortex distinctly paraplectenchymatous, colourless; apical cells at the lobate region greyish black.

Apothecia rare, adnate, up to 1 mm across. Disc at first plane, later somewhat convex, black, epruinose. Exciple black, thin, evanescent. Hymenium colourless, 60–110 μ thick. Hypothecium dark brown, 40–50 μ thick. Epithecium brownish black. Paraphyses discrete, septate and branched; apical cells enlarged and black. Spores 8, brown, 1-septate, ellipsoid, 9–12 × 6–7.5 μ.

Reactions: Thallus K+ deep yellow, C–; medulla Pd+ yellow; contains chloroatranorin (Pl. XXVII: 1); hymenium I+ blue.

Habitat: On flint. W. and C. Negev. On olive and fig trees and on *Rhamnus*. Upper Galilee, Shefela, Samaria.

Distribution: Euro-Siberian, Mediterranean, Irano-Turanian and Saharo-Arabian; also found in N. America.

3. PHYSCIA (Schreb.) DC.

Thallus foliose, whitish to pale grey or brown, dorsiventral, appressed or ascending; rhizinate. Upper cortex paraplectenchymatous, K+ yellow (atranorin) or rarely K–. Lower cortex plectenchymatous, hyaline or externally pigmented. Phycobiont *Trebouxia* (Fig. 1 A).

Apothecia lecanorine. Spores (Fig. 13 B) *Physcia* type (Poelt, 1965), 1-septate, at first greenish, later brown; cell walls strongly thickened at the apices and septum; surface smooth.

1. Lobes with marginal and apical cilia — 2
- Lobes without cilia — 4
2. Thallus sorediate — 3
- Thallus without soralia. — **6. P. leptalea**
3. Lobes with inflated cup-shaped sorediate tips. — **4. P. ascendens**
- Lobes with apical labriform soralia. — **5. P. tenella**
4. Cortex and medulla K+ yellow. Upper surface with white spots. — **1. P. aipolia**
- Cortex K+ yellow, medulla K–. Upper surface without white spots, or white spots inconspicuous — 5
5. Thallus densely pruinose, without white spots. Lower surface yellowish at the centre. — **2. P. biziana**
- Thallus epruinose or slightly pruinose, with white spots faintly visible when wet. Lower surface whitish. — **3. P. stellaris**

1. Physcia aipolia (Ehrh.) Hampe, in Fürnr., Naturh. Topogr. Regenburg 2: 249 (1839). *Lichen aipolius* Ehrh., in Humb., Fl. Friburg. Spec. 19 (1793).

Thallus foliose, rather closely attached to the substrate, rosulate; rosettes 4–6 (–8) cm across; lobes thick and rigid, long and narrow, pressed against each other or separate, plane at the periphery, convex and conglomerate at the centre; tips broadened, crenulate; upper surface greyish white or ash-grey, epruinose, densely covered with irregular white spots (pseudocyphellae; Maas Geesteranus, 1952), at first smooth, later becoming rough and wrinkled; lower side dull white or brownish in places, densely covered by whitish or brown rhizines. Upper cortex 45–75 μ thick, paraplectenchymatous, colourless, with a yellowish brown amorphous layer on top and interrupted by algal clusters and medullary tissue causing the white spots on the surface. Medulla white. Lower cortex plectenchymatous, a more densely aggregated continuation of medullary hyphae, colourless with an exterior dark brown zone.

Apothecia numerous, centrally located, adnate, up to 2 mm across. Disc plane, brown or black, heavily pruinose. Thalline margin thick, persistent, at first entire, later crenulate. Hymenium colourless, 90–120 μ thick. Epithecium yellowish brown. Hypothecium colourless, 70–80 μ thick. Spores 8, dark brown, 2-celled, ellipsoid, straight or somewhat curved, cell walls strongly thickened at the septum and at the rounded ends, 19–25 × 9–11 μ.

Reactions: Cortex and medulla K+ yellow; contains atranorin (Pl. XXV: 1); hymenium I+ blue turning reddish.

Habitat: On bark of olive trees. Upper Galilee.

Distribution: Common in the N. Hemisphere.

2. Physcia biziana (Mass.) Zahlbr., Oester. Bot. Z. 51 : 26 (1901). *Squamaria biziana* Mass., Miscell. Lich. 35 (1856). [Plate XVII: 3]

Thallus foliose, forming rosettes up to 5 cm across, closely appressed to the substrate; lobes flat or moderately convex, profusely branched, slightly apart from each other or approaching but not overlapping; margins somewhat crenate and incised; tips bent downward; centre sometimes wart-like; upper surface chalk-white, strongly pruinose; lower surface whitish along the periphery, yellowish in the centre, with short rhizines of the same colour. Upper cortex 25–45 μ thick, colourless, paraplectenchymatous. Algal layer more or less continuous. Medulla white. Lower cortex formed of hyphae parallel to the surface, or nearly so, and not clearly separated from the medulla.

Apothecia usually crowded in the centre, 1–3 mm across, sessile or somewhat pedicillate. Disc concave to plane or convex, black, naked or densely pruinose. Thalline margin thick and involute, persistent, entire or occasionally crenulate. Hymenium 70–85 μ thick, colourless. Epithecium brown, 20–30 μ thick. Hypothecium colourless or faintly yellow below, 55–75 μ thick. Spores 8, greenish brown or brown, 2-celled, ellipsoid or sometimes beanlike with thick walls especially at the ends and centre, 15–18 × 6–8 μ.

Reactions: Cortex K+ yellow, medulla K–; contains atranorin (Pl. XXV: 1); hymenium I+ blue.

Habitat: On bark of *Casuarina*. Upper Jordan Valley.

Distribution: Mediterranean.

3. Physcia stellaris (L.) Nyl., Act. Soc. Linn. Bordeaux 21 : 307 (1856). *Lichen stellaris* L., Spec. Plant. 1144 (1753). [Plate XVII: 1]

Thallus foliose, forming more or less stellate rosettes 1–5 cm across in rather close contact with the substrate; lobes 0.8–1.2 cm long and 1–1.5 mm broad, with somewhat broader tips, rigid, discrete or contiguous, richly and irregularly branched; centre wrinkled and warty; upper surface ash-grey, epruinose or slightly pruinose, white spots only faintly visible in wet plants; lower side dirty white with pale or dark-tipped rhizines sometimes projecting sideways. Upper cortex paraplectenchymatous, colourless, covered by an amorphous brownish exterior zone. Algae, *Trebouxia impressa* (Ahmadjian, 1960), in clusters, sometimes reaching close to the upper surface. Medulla white, merging into the lower cortex, composed of hyphae more or less parallel to the surface, with an exterior brown zone.

Apothecia numerous, sessile or shortly pedicellate, 1–3 mm across. Disc plane, dark brown or black, usually pruinose. Thalline margin persistent, prominent, entire or crenulate. Hymenium 80–90 μ thick, colourless with a yellow brown epithecium. Hypothecium yellowish, 25–40 μ thick. Spores 8, 2-celled, brown when mature, ellipsoid, straight or curved, with strongly thickened cell walls at the septum and rounded apices, 16–24 × 7–10 μ.

Reactions: Cortex K+ yellow; medulla K–; contains atranorin (Pl. XXV: 1); hymenium I+ blue turning wine-red.

Habitat: On oak and olive trees. Upper Galilee.

Distribution: In temperate regions of the N. Hemisphere.

4. Physcia ascendens Bitt., in Pringsh. Jahrb. Wiss. Bot. 36: 431 (1901). [Plate XVII: 2]

Thallus foliose; individual thalli not more than 3 cm across, but several often fusing to cover extensive areas; lobes long and narrow (1–1.5 cm long, 0.5–1 mm broad), ascending, irregularly rebranched and discrete, plane or convex, loosely attached to the substrate by cilia and rhizines; cilia 1–3 mm long, simple with brown tips or brown all over, usually marginal or apical, some emergent from the lower side of the lobes and partly serving as hapters, the rest projecting into the air; rhizines sparse, pale; upper side light greyish white, sometimes pruinose; lower side whitish; lobe tips sorediate. Soralia cup- or helmet-shaped, turning upside down or sideways, formed by a break between the medulla and the upper cortex and the inflation of the latter; with increase of soredial production some deformations occur. Upper cortex of more or less perpendicular colourless cell rows covered by an amorphous brownish layer. Algae in an irregular layer penetrating both cortex and medulla. Medulla white, gradually merging into lower cortex; lower cortex composed of parallel hyphae, exterior zone brown in places.

Apothecia very rare, sessile or pedicellate, up to 2 mm across. Disc concave to plane, dark brown, pruinose. Thalline margin entire, prominent, persistent. Hymenium colourless, 65–85 μ thick. Epithecium brownish. Hypothecium colourless. Spores 8, 2-celled, olive-brown, ellipsoid, straight or slightly curved, with thickened cell walls at the septum and rounded apices, 15–18 × 5–8 μ.

Reactions: Cortex K+ yellow, Pd+ yellow; medulla K–; contains atranorin (Pl. XXV: 1); hymenium I+ blue.

Habitat: On dolomite, basalt and twigs of *Rhamnus palaestina*. Upper Galilee, Upper Jordan Valley, Coast of Carmel, Shefela, Judean Mountains.

Distribution: Common in the N. Hemisphere.

5. Physcia tenella (Scop.) Bitt., in Pringsh. Jahrb. Wiss. Bot. 36: 431 (1901). *Lichen tenellus* Scop., Fl., Carniol. ed. 2, 2: 394 (1772).

Very similar to *P. ascendens;* rosettes 1–2 cm across; several thalli sometimes fusing; lobes at first adnate, later slightly ascending, up to 1 cm long and 0.2–1 mm broad, discrete, overlapping or entangled, ciliate; cilia marginal and apical, simple, 1–2 mm long, brown all over or with brown tips; upper surface light greyish white, sometimes pruinose, sorediate at the tips of the lobes; lower surface whitish with few, pale rhizines. Soralia labriform, formed by a break between the medulla and the cortex and the reflexing of the latter, finally appearing as large clusters of soralia on the lobes. Thallus layers, reactions and apothecia as in *P. ascendens*.

Habitat: On olive trees and calcareous rocks. Upper Galilee, Shefela, Judean Mountains.

Distribution: N. Hemisphere (less common than *P. ascendens*).

6. **Physcia leptalea** (Ach.) DC., in Lam. and DC., Fl. Fr. ed. 3, 2: 395 (1805). *Lichen leptaleus* Ach., Lich. Suec. Prod. 108 (1798). [Fig. 14]

Thallus foliose, loosely attached to the substrate, forming rosettes 2–4 cm across; often several thalli coalescent; lobes more or less stellate, *c.* 1 cm long and 0.2–0.8 mm broad, discrete to entangled, richly and irregularly branched, plane or convex, ciliate; cilia marginal and apical, simple, 1–2 mm long, pale or with brown tips; upper surface pale ash-grey with somewhat conspicuous white spots (pseudocyphellae); lower surface dull, white or brownish in places with a few pale rhizines. Upper cortex of more or less perpendicularly arranged cell rows, colourless; exterior zone amorphous brown. Algae in clusters of which some almost reach the upper surface. Medulla white, emerging into a lower cortex of densely arranged more or less longitudinal hyphae; exterior brownish.

Apothecia numerous, laminal and marginal, sessile or shortly stalked, 2–4 mm across. Disc concave or plane, dark brown, usually pruinose. Thalline margin persistent, prominent, entire or crenulate. Hymenium colourless. Epithecium yellowish brown. Hypothecium yellowish. Spores 8, 2-celled, brown when mature, ellipsoid, straight or curved, with strongly thickened cell walls at the septum and rounded ends, 15–22 × 8–10 μ.

Reactions: Cortex K+ yellow; medulla K−; contains atranorin (Pl. XXV: 1); hymenium I+ blue.

Habitat: On oak and olive trees. Upper Galilee.

Distribution: Medio-European and N. Mediterranean; also found in N. America.

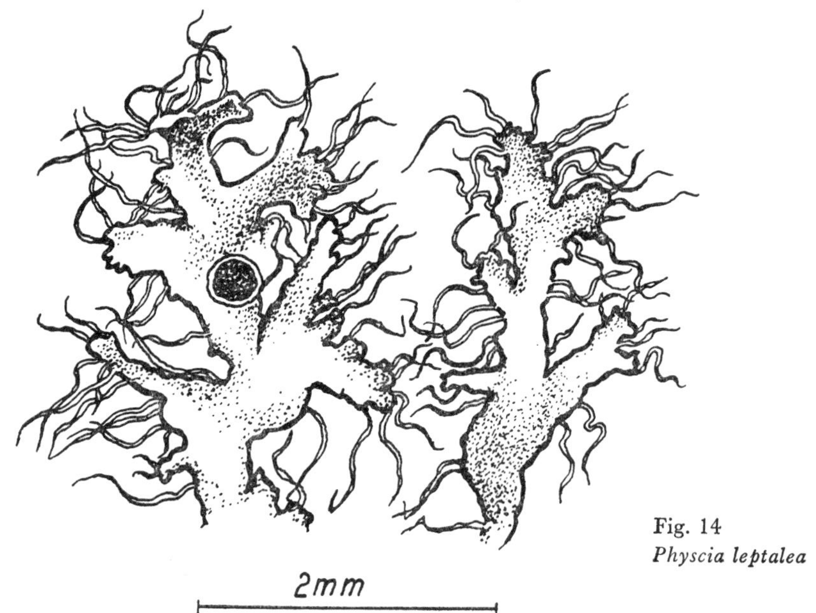

Fig. 14
Physcia leptalea

4. PHYSCONIA Poelt

Thallus foliose, grey to brown, dorsiventral, rhizinate. Upper cortex proso- or paraplectenchymatous, K–. Phycobiont *Trebouxia* (Fig. 1 A). Medulla white or yellowish; lower surface of medullary hyphae aggregated in a more or less longitudinal direction, pale or black.

Apothecia lecanorine. Spores (Fig. 13 C) (*Physconia* type; Poelt, 1965), 1-septate, at first greenish, later brown; surface sculptured with tiny warts; external cell walls moderately thickened, very thick at the septum, with a broad canal connecting the lumina.

1. Cortex paraplectenchymatous. Rhizines dichotomously branched (Fig. 15 A). **1. P. grisea**
– Cortex prosoplectenchymatous. Rhizines squarrosely branched (Fig. 15 B) 2
2. Thallus sorediate. **5. P. farrea**
– Thallus without soredia 3
3. Medulla white, K– 4
– Medulla yellow, more so with K. **3. P. subpulverulenta**
4. Lower surface black, except at the growing tips. **2. P. pulverulenta**
– Lower surface whitish. **4. P. venusta**

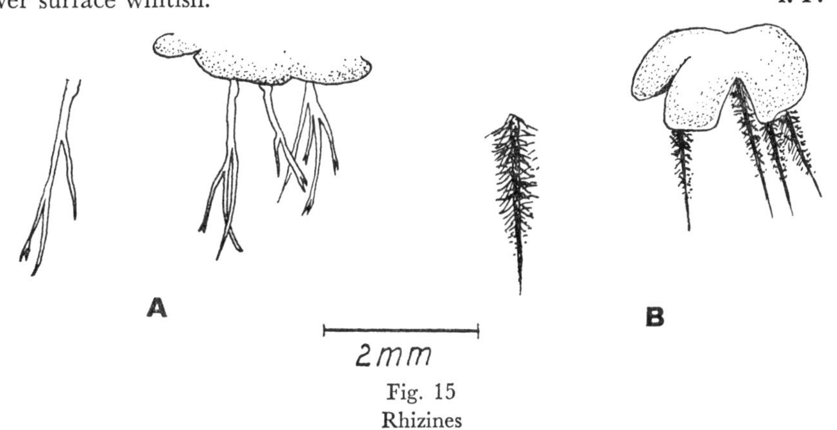

Fig. 15
Rhizines
A. dichotomously branched; B. squarrosely branched

1. Physconia grisea (Lam.) Poelt ssp. *lilacina* (Arn.) Poelt, Nova Hedwigia 12 : 120 (1966). *Parmelia pulverulenta* f. *lilacina* Arn., Flora 46 : 589 (1863).

Thallus foliose, usually forming rosettes or spreading, covering large areas, attached fairly closely to the substrate, 140–300 μ thick; lobes broad, up to 5 mm at the periphery, adjacent or imbricate; centre occupied by numerous imbricate simple or divided secondary laciniae; laciniae marginal, flat, usually oriented in one direction; upper surface grey or greyish brown, usually pruinose, sometimes only the laciniae covered with pruina; pruina bluish lilac; isidia and soredia absent; lower surface pale; rhizines abundant, at first pale, later dark at the tips, dichotomously branched (Fig. 15 A). Upper cortex paraplectenchymatous, uppermost cells brownish, usually covered by a colourless amorphous stratum. Medulla and lower cortex white.

Sterile.

Reactions: All reactions negative; contains no lichen acids.

Habitat: Abundant on *Pistacia,* on mossy soil, on mossy and bare calcareous rocks. Upper and Lower Galilee, Judean Mountains.

Distribution: Medio-European and Mediterranean.

2. Physconia pulverulenta (Schreb.) Poelt, Nova Hedwigia 9: 30 (1965). *Lichen pulverulentus* Schreb., Spic. Fl. Lips. 128 (1771). [Plate XVIII: 3]

Thallus foliose, externally very polymorphic, orbicular or irregularly widespread, covering large areas, fairly closely attached to the substrate, 200–250 μ thick; lobes long and narrow, 5–12 mm long and 0.5–2 mm broad, discrete or entangled, contiguous or partly overlapping, usually flat; upper side grey, greyish brown, or brown, usually pruinose, the pattern of the pruina very variable in different specimens; isidia and soredia absent; lower side blackish, except at the growing tips; densely covered by black rhizines; rhizines squarrosely branched (Fig. 15 B). Upper cortex prosoplectenchymatous, colourless; uppermost cells brownish, usually with an amorphous stratum above. Medulla white. Lower cortex blackish.

Apothecia absent or richly developed, 1–3 mm across, laminal. Disc dark brown or black, naked or pruinose or both on the same thallus. Thalline margin concolourous with the thallus, persistent, entire or warted, often with a corona of lobules. Hymenium colourless with a brownish gelatinous epithecium. Hypothecium yellowish, covering an algal layer. Paraphyses septate and furcate. Spores 8, at first greenish, later brown, 2-celled with broadly rounded ends, straight or occasionally curved, with walls thickened, especially at the septum, 18–38 × 8–20 μ.

Reactions: All reactions negative; contains no lichen acids.

Habitat: Common on bark of *Pistacia, Olea, Rhamnus* and on calcareous rocks. Upper Galilee.

Distribution: Medio-European, Mediterranean and Irano-Anatolian; also found in N. America.

3. Physconia subpulverulenta (Szat.) Poelt, Nova Hedwigia 12: 127 (1966). *Physcia subpulverulenta* Szat., Bortasia 3: 135 (1941).

Very similar to *P. pulverulenta,* but with more lobules in the centre and from the thalline margin of the apothecia. Medulla yellowish, turning clear yellow with K.

Habitat: On calcareous rocks and on *Quercus calliprinos.* Upper Galilee.

Distribution: Mediterranean.

4. Physconia venusta (Ach.) Poelt, Nova Hedwigia 12: 130 (1960). *Parmelia venusta* Ach., Method. Lich. 211 (1803). [Plate XVIII: 2]

Differs from *P. pulverulenta* in its whitish lower surface and the richer development of adventive lobules, especially around the apothecia.

Habitat: On bark of *Casuarina.* Rare. Upper Galilee.

Distribution: Mediterranean.

5. Physconia farrea (Ach.) Poelt, Nova Hedwigia 9 : 30 (1965). *Parmelia farrea* Ach., Lichgr. Univ. 475 (1810). [Plate XVIII : 1]

Thallus foliose, at first orbicular, later irregularly widespread, 3–8 cm across, 160–250 μ thick, fairly closely attached to the substrate; lobes rigid, contiguous or partly overlapping at the periphery, imbricate or without any order in the centre, at first plane, later concave with slightly ascending margins; margins densely sorediate, soredia labriform; upper surface greyish brown or greyish green, pruinose; pruina whitish or bluish white, usually covering 2/3 of the lobes; lower surface whitish, obscure in places, usually around the rhizines; rhizines black, squarrosely branched (Fig. 15 B). Upper cortex 30–60 μ thick, composed of irregularly arranged hyphae with prosoplectenchymatous cells, colourless, sometimes with a gelatinous yellowish layer above. Medulla white. Lower cortex indistinct, limited by medullary edge cells. Sterile.

Reactions : All reactions negative; contains no lichen acids.

Habitat: On *Quercus calliprinos, Rhamnus palaestina, Olea europea, Pistacia palaestina.* Upper Galilee.

Distribution : Medio-European, Mediterranean and Irano-Anatolian.

5. ANAPTYCHIA Körb.

Thallus foliose, dorsiventral, lobes linear-elongate, corticate on the upper side only or on both sides; cortical hyphae oriented in a longitudinal direction more or less parallel to the surface. Apothecia lecanorine, sessile or somewhat stalked. Spores 8, brown, ellipsoid, 2-celled, with thin or thickened cell walls.

1. Anaptychia ciliaris (L.) Körb., in Mass. Mem. Lichgr. 35 (1853). *Lichen ciliaris* L., Sp. Pl. 2 : 1144 (1753). [Fig. 16]

Thallus foliose, covering extensive areas, up to or exceeding 20 cm across, laciniate; laciniae repeatedly dichotomising, linear-elongate, 1–2 mm broad, approx. 300 μ thick, apices slightly ascending but generally decumbent; cilia marginal, 1–6 mm long, simple or rarely branched, concolourous with the thallus or darker, tomentose, some serving as hapters; upper surface greyish white, tomentose; lower surface paler. Cortex irregular, composed of thick-walled hyphae parallel to the surface, sometimes filling out an entire section down to the lower surface. Algae in clusters. Medulla very thin or lacking. Lower cortex absent.

Apothecia laminal, stipitate or subsessile, 2–4 mm across. Disc brown, pruinose or naked. Apothecia when young surrounded by an entire thalline margin which later becomes distinctly lacinulate. Hymenium colourless, 150–200 μ thick. Epithecium yellowish brown. Hypothecium colourless or faintly yellowish. Spores 8, dark brown, 2-celled, ellipsoid with rounded ends, slightly constricted at the centre with uniform cell wall, 25–45 × 15–25 μ.

Reactions : Thallus and medulla K–, C–, KC–, Pd–; contains no lichen acids.

Habitat: On trunks and branches of *Quercus calliprinos, Pistacia palaestina* and *Rhamnus palaestina.* Upper Galilee.

Distribution : Common in the N. Hemisphere.

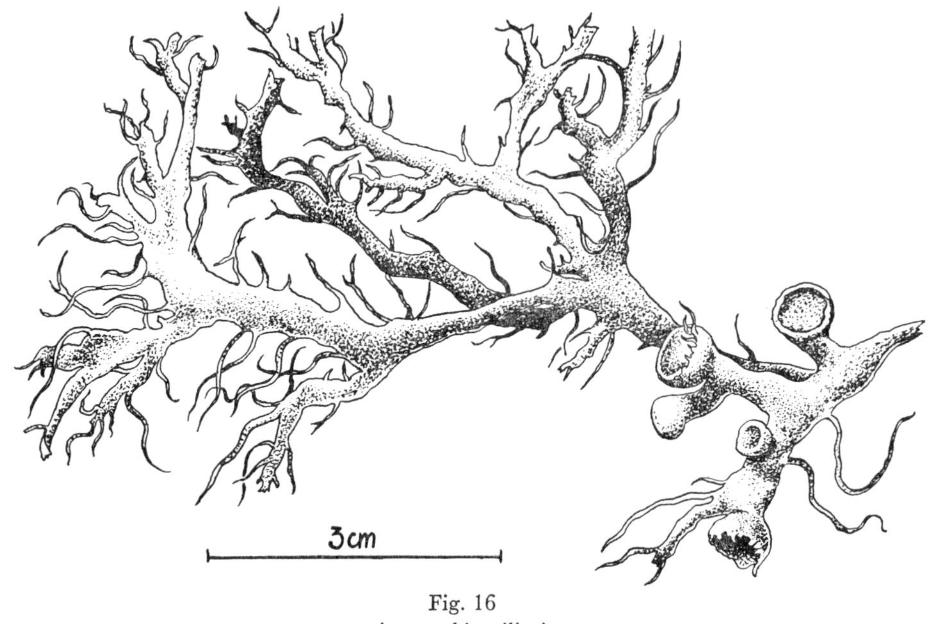

Fig. 16
Anaptychia ciliaris

6. Tornabenia Mass.

Thallus fruticose; branches more or less radial, corticate. Cortex of thick-walled longitudinal hyphae. Apothecia lecanorine, sessile or substipitate. Spores 8, brown, placodiomorph.

Tornabenia intricata Trevis., Tornabenia et Blasteniaspora 1 (1853). *Anaptychia intricata* (Desfont.) Mass. [Plate XVIII : 4]

Thallus fruticose, growing in laciniate tufts or clumps around tree branches, up to or exceeding 10 cm in height and width; laciniae very densely intricate and irregularly branched, narrow (0.5–1.5 mm); primary laciniae more or less compressed; secondary laciniae almost cylindrical or fibrilliform; surface grey or greyish brown with a velvety cover, some parts glabrous. Cortical hyphae colourless, parallel to the surface. Algal layer thick and continuous on one side of the medulla, thinner and interrupted at the other. Medulla loose, colourless.

Apothecia numerous, lateral, scattered, 0.5–2 mm across. Disc brown or blackish, naked or slightly pruinose, plane. Thalline margin entire, prominent and persistent. Hymenium and hypothecium colourless. Epithecium dark brown. Spores 8, brown, placodiomorph, 20–26 × 9–15 μ.

Reactions: All reactions negative; contains no lichen acids.

Habitat: Abundant on *Rhamnus palaestina,* also on *Pistacia palaestina* and *Quercus calliprinos.* Upper Galilee.

Distribution: Medio-European and Mediterranean; also found rarely in the British Isles.

TELOSCHISTACEAE

Thallus crustose, effuse or lobate-effigurate, foliose or fruticose, stratified, yellow or orange and K+ purple, or white, grey, brown or blackish and K–. Cortex above, on both sides or undefined. Phycobiont belonging to the Chlorophyceae (mostly *Trebouxia*) (Fig. 1 A). Apothecia lecanorine or biatorine. Epithecium K+ purple or violet. Paraphyses septate, simple or furcate. Spores usually 8, colourless, placodiomorph or simple, rarely with a thin septum (*Fulgensia*), then thallus orange and K+ purple.

1.	Thallus crustose, effuse, determinate or lobate-effigurate	2
–	Not as above	4
2.	Spores thin-walled, simple or with one thin septum.	**3. Fulgensia**
–	Spores placodiomorph	3
3.	Apothecia lecanorine.	**2. Caloplaca**
–	Apothecia biatorine.	**1. Blastenia**
4.	Thallus foliose. Cortical hyphae perpendicular to the surface.	**5. Xanthoria**
–	Thallus fruticose. Cortical hyphae longitudinal.	**4. Teloschistes**

1. BLASTENIA Mass.

Thallus crustose. Apothecia immersed or superficial, biatorine. Spores, 4–16, colourless, placodiomorph.

1.	Thallus golden to ochreous yellow. Apothecia orange.	**1. B. latzeli**
–	Thallus white or greyish white. Apothecia black.	**2. B. rejecta** var. **bicolor**

1. Blastenia latzeli Serv., Hedwigia 74 : 151 (1934). *Caloplaca dalmatica* (Mass.) Zahlbr.

Thallus at first orbicular, later indeterminate, very thin, continuous or faintly areolate, golden to ochreous yellow and whitish-pruinose.

Apothecia minute, up to 0.5 mm across, biatorine, immersed or adnate. Disc dark orange, plane. Exciple paler than the disc, persistent, 50–65 μ thick. Hypothecium yellowish. Epithecium yellow-orange, granular. Spores 8, colourless, ellipsoid, 12.5–18 × 8–10 μ, mischoblastiomorph (Pl. XIX : 1).

Reactions : Thallus and apothecia K+ purple; contains parietin.

Habitat : On calcareous rocks. Lower Galilee, Judean Mountains.

Distribution : Reported once from Yugoslavia.

2. Blastenia rejecta Th. Fr. var. **bicolor** (Müll. Arg.) Zahlbr., Cat. Lichgr. Univ. 7 : 41 (1931). *B. melanocarpa* var. *bicolor* Müll. Arg., Rev. Myc. 2 : 78 (1880). [Plate XIX : 2]

Thallus more or less orbicular, 1–3 cm across, solitary or several confluent, white or greyish white, farinose, 150–400 μ thick, smooth or rimose-areolate especially towards the periphery.

Apothecia numerous, mainly concentrated in the centre, 0.8–1 mm across, at first somewhat immersed, later adnate on slightly elevated areoles. Disc approximately plane, black, epruinose. Exciple dark brown; internal cellular tissue colourless, surrounded by a thin brown cortex. Algal layer continuous beneath the hypothecium, laterally as high as the hypothecium or somewhat higher. Hymenium colourless, 100–150 μ thick. Hypothecium colourless. Paraphyses septate, gradually broadening above; apices dark brown. Spores 8, colourless, ellipsoid, placodiomorph, 13.5–16 × 6–7.5 μ, isthmus 3–4.5 μ.

Reactions: Thallus K−; epithecium K+ violet; hymenium I+ blue; hypothecium I+ violet.

Habitat: On calcareous rocks. C. Negev.

Distribution: Saharo-Arabian.

2. Caloplaca Th. Fr.

Thallus crustose, without a clearly defined cortex, or lobate-effigurate to squamulose with paraplectenchymatous cortex, closely attached to the substrate, without rhizines, yellow to orange-red and K+ purple, or white, grey or blackish and usually K−. Phycobiont *Trebouxia* (Fig. 1 A). Apothecia lecanorine or rarely biatorine, sessile or rarely innate. Hypothecium colourless, subtended by algal cells. Spores 8, colourless, usually placodiomorph or becoming 3–4-celled with connecting isthmus.

Section caloplaca Th. Fr. Thallus crustose. Cortex undifferentiated. Spores placodiomorph.

1. Thallus sorediate. **2. C. citrina**
– Thallus esorediate 2
2. Thallus shades of yellow or yellowish. **1. C. velana**
– Thallus greyish or whitish 3
3. Apothecia dark brown to black 4
– Apothecia shades of orange or red 5
4. Thallus K± yellow. Thalline margin disappearing. Spores subglobose. **6. C. interveniens**
– Thallus K+ violet. Thalline margin persistent. Spores ellipsoid. **9. C. variabilis**
5. Corticolous 6
– Saxicolous 7
6. Hypothallus bluish black. Apothecia reddish brown. **5. C. haematites**
– Hypothallus absent. Apothecia bright orange. **7. C. luteoalba**
7. On basalt. Hypothallus blackish blue. Apothecia rust-red. **3. C. festiva**
– On calcareous rocks. Hypothallus absent. Apothecia orange to brownish 8
8. Spores placodiomorph; isthmus about 1/5 of the spore length. **8. C. lamprocheila**
– Spores with a thin septum, rarely with a very short isthmus. **4. C. flageyana**

1. Caloplaca velana DR., Gotl. Vegetationsstud. 45 (1925).

Thallus rather determinate, firmly attached to the substrate, *c.* 3 cm across, 1–1.5 mm thick, dull yellow, cracked-areolate.

Apothecia numerous, sessile, up to 0.8 mm across. Disc deep to brownish orange, plane or somewhat convex. Thalline margin thin, entire or flexuous, disappearing or persistent. Exciple thicker, concolourous with the disc or somewhat paler, entire or crenulate, prominent and persistent, sometimes glossy. Hymenium colourless, 60–80 μ thick. Epithecium yellow, granular. Hypothecium colourless, up to 110 μ thick. Exciple 30–40 μ thick below, up to 60 μ thick above. Paraphyses simple or furcate. Spores 8, colourless, placodiomorph, broadly ellipsoid, 10–12 × 4.5–8 μ; isthmus about 1/3 as long as the spore.

Reactions: Thallus and apothecia K+ purple; contains parietin (Pl. XXVIII: 3).

Habitat: On basalt. Esdraelon Plain.

Distribution: Cosmopolitan.

2. **Caloplaca citrina** (Hoffm.) Th. Fr., Nova Acta Soc. Sci. Upsal., ser. 3, 3: 218 (1861). *Verrucaria citrina* Hoffm., Deutsch. Flora 198 (1796).

Thallus a bright lemon-yellow crust, at first scaly, finally completely granular-sorediate.

Apothecia rare, sessile, *c.* 0.5 mm across. Disc plane or slightly convex, orange-yellow. Thalline margin at first entire, later sorediose and usually disappearing. Exciple somewhat paler than the disc, entire. Hymenium colourless, 50–60 μ thick. Epithecium granular, yellowish. Paraphyses furcate; apices clavate. Spores 8, ellipsoid, colourless, placodiomorph, 10–15 × 5–7 μ; isthmus about 1/3 as long as the spore.

Reactions: Thallus and apothecia K+ purple; contains parietin (Pl. XXVIII: 3).

Habitat: On calcareous rocks, on mortar of walls and among mosses and *Squamarina crassa* squamules. Mt. Carmel, Shefela.

3. **Caloplaca festiva** (E. Fr.) Zw., Flora 47: 85 (1864). *Biatora ferruginea* var. *festiva* E. Fr., Nov. Sched. Critic. 8 (1827). [Plate XIX: 6]

Thallus effuse or determinate, dark grey, thin, areolate, on a bluish black hypothallus.

Apothecia numerous, sessile, 0.2–1.5 mm across. Disc plane, rust-red. Thalline margin concolourous with the disc or somewhat darker, glossy, rather thick, at first prominent, later level with the disc. Exciple 20–30 μ at the base, up to 80–100 μ above, dark orange, persistent. Hymenium and hypothecium colourless. Epithecium dark orange, granular. Spores 8, ellipsoid, colourless, placodiomorph, 13–15 × 7.5–8.5 μ; isthmus usually 5 μ.

Reactions: Thallus K–; apothecia K+ violet; hymenium and hypothecium I+ blue; exciple I–.

Habitat: On basalt. Upper Jordan Valley.

Distribution: Submediterranean.

4. **Caloplaca flageyana** (Flag.) Zahlbr., Cat. Lichgr. Univ. 7: 130 (1931). *Gyalolechia cinnabarina* Flag., Rev. Myc. 17: 104 (1895).

Thallus very thin, grey, almost invisible, warty or partly rimose, in small (1.0–1.5 cm) roundish patches; several sometimes fused.

Apothecia many, more or less uniformly dispersed, sessile, up to 0.5 mm across, biatorine. Disc rufous orange, at first concave, later plane. Margin somewhat paler than the disc, at first thick and inflated, later partly rolled back. Hymenium colourless, 90–100 μ thick. Epithecium granular, orange. Hypothecium colourless, 40–60 μ thick. Paraphyses simple, rarely furcate; 1–4 upper cells short and thickened. Spores 8, colourless, ellipsoid, mostly 1-septate, more or less uniform in size, 18 × 6–7.5 μ; septum as thick as the spore wall, some placodiomorph with large lumina and a very short isthmus (Fig. 17 A).

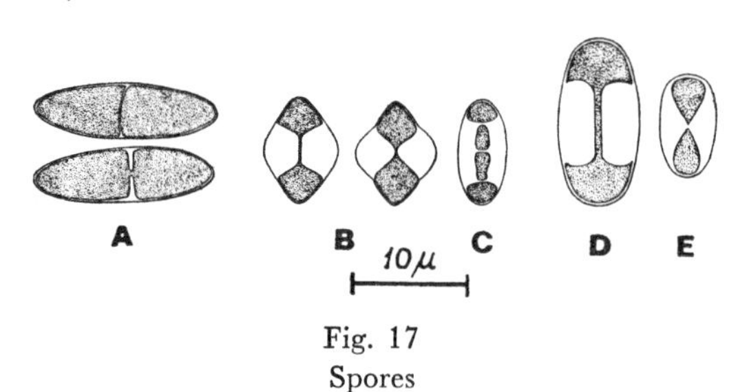

Fig. 17
Spores
A. *Caloplaca flageyana*; B. *Caloplaca aurantia* var. *aurantia*;
C. *Caloplaca ochracea*; D. *Xanthoria parietina*; E. *Xanthoria steineri*

Reactions: Thallus K–; apothecia K+ purple; contains parietin (Pl. XXXIII : 3).
Habitat: On dolomite and limestone. C. Negev.
Distribution: Reported only from Algeria.

5. Caloplaca haematites (Chaub.) Zw., Flora 45 : 487 (1862). *Lecanora haematites* Chaub., in Saint-Amans, Fl. Agenaise 492 (1821).

Thallus determinate, 1–2 cm across, *c.* 100 μ thick, dark grey; centre composed of irregular, plane, very small areoles; circumference granular. Hypothallus bluish black.

Apothecia crowded in the centre, sessile, up to 0.6 mm across. Disc reddish brown, at first concave, later plane. Thalline margin concolourous with the thallus, entire, persistent. Exciple concolourous with the disc, 5–15 μ at the base, 30–35 μ above. Hymenium colourless, 60–75 μ thick. Epithecium orange, granular. Hypothecium colourless, 45–75 μ thick. Paraphyses simple or rarely branched, gradually thickening towards the ends. Spores 8, colourless, placodiomorph, subglobose or ellipsoid, 10.5–12 × 7.5–9 μ; isthmus 6–7.5 μ long.
Reactions: Thallus K+ yellowish; apothecia K+ purple.
Habitat: On olive trees. Esdraelon Plain.
Distribution: Medio-European, Atlantic and Mediterranean.

6. Caloplaca interveniens (Müll. Arg.) Zahlbr., Cat. Lichgr. Univ. 7 : 146 (1931). *Callopisma interveniens* Müll. Arg., Rev. Myc. 6 : 17 (1884).

Thallus more or less determinate, 1–2 cm across, grey with a brown or reddish tinge, solitary or several contiguous and delimited from each other by a dark grey line.

Apothecia numerous, up to 0.7 mm across, innate, level with the thallus or adnate.

Disc black, naked or pruinose, plane or somewhat convex. Thalline margin concolourous with the thallus, disappearing. Hymenium colourless, 90–140 μ thick. Hypothecium colourless, 60–70 μ thick. Epithecium obscure, granular. Paraphyses simple, clavate. Spores 8, colourless, placodiomorph, subglobose, 10.5–13.5 × 7.5–9 μ; isthmus about half as long as the spore.

Reactions: Thallus K+ faintly yellow; epithecium K+ violet.

Habitat: On limestone and crystalline dolomite. N. Negev.

Distribution: Reported only from Egypt.

7. **Caloplaca luteoalba** (Turn.) Th. Fr., Nov. Acta Soc. Sci. Upsal. ser. 3, 3: 220 (1861). *Lichen luteoalbus* Turn., Trans. Linn. Soc. London 7: 92 (1803). [Plate XIX: 4]

Thallus effuse, tartareous, light greyish brown, 100–160 μ thick, almost invisible.

Apothecia numerous and crowded, up to 0.6 mm across. Disc bright orange, at first plane, later convex. Thalline margin thick, entire, somewhat paler than the disc, at first prominent, finally almost excluded. Hymenium colourless, 70–80 μ thick. Epithecium of glossy orange granules. Hypothecium colourless, 40–60 μ thick, subtended by a continuous algal layer. Paraphyses simple or furcate, capitate. Spores 8, ellipsoid, colourless, placodiomorph, 8–12 × 3–6 μ; isthmus about 1/5 as long as the spore.

Reactions: Thallus K–; apothecia K+ purple; contain parietin (Pl. XXVIII: 3).

Habitat: On *Acacia tortilis*. C. Negev.

Distribution: Pluriregional.

8. **Caloplaca lamprocheila** Flag., Rev. Myc. 10: 130 (1880).

Thallus scarcely visible, tartareous, whitish grey.

Apothecia numerous, sessile, up to 0.5 mm across, biatorine. Disc chestnut-brown, at first plane, later somewhat convex. Margin concolourous with the disc, thin, persistent. Hymenium colourless, 90–100 μ thick. Epithecium orange, granular. Hypothecium colourless, 45–65 μ thick. Paraphyses simple or furcate; 1–4 uppermost cells short and thickened. Spores 8, colourless, ellipsoid, placodiomorph, 13.5–15 × 4.5–7 μ; isthmus 1/5 as long as the spore.

Reactions: Thallus K–; apothecia K+ purple.

Habitat: On dolomite, limestone and basalt. Upper Galilee, Mt. Gilboa, Upper Jordan Valley, Coast of Carmel, Sharon Plain, C. Negev.

9. **Caloplaca variabilis** (Pers.) Müll. Arg., Mém. Soc. Phys. et Hist. Nat. Genève 16: 387 (1862). *Lichen variabilis* Pers., in Usteri Neue Ann. Bot. 1: 26 (1794).

Thallus more or less determinate, greyish brown to almost black, areolate; areoles 0.2–0.8 mm across, plane to moderately convex. Hypothallus black, not always present.

Apothecia numerous, partly immersed or superficial, up to 0.8 mm across. Disc dark brown or black, pruinose, plane or convex. Thalline margin entire, persistent, whitish, pruinose. Hymenium colourless, 60–100 μ thick. Epithecium olive-brown, granular. Hypothecium colourless. Paraphyses simple or furcate, clavate. Spores 8, placodiomorph, ellipsoid, 11–16 × 7–9 μ; isthmus 1/3 as long as the spore.

Reactions: Thallus and epithecium K+ violet.

Habitat: On calcareous rocks. Upper and Lower Galilee, Mt. Carmel.

Distribution: Common in temperate regions of the N. Hemisphere.

Section GASPARRINA (Tornab.) Th. Fr. Thallus lobate-effigurate or squamulose. Cortex paraplectenchymatous. Spores placodiomorph.

1. Thallus yellow-orange to orange, K+ purple — 2
- Thallus of a different colour, K– or K+ — 3
2. Spores ellipsoid. — **16. C. murorum**
- Spores lemon-shaped to rhomboid. — 11. C. **aurantia** var. **aurantia**
3. Thallus poorly developed, sorediate. — **17. C. negevensis**
- Thallus well developed, esorediate — 4
4. Thallus squamulose-lobate, blackish grey. — **18. C. conglomerata**
- Thallus crustose-effigurate, paler — 5
5. Centre and periphery concolourous, K– or K+ yellow — 6
- Centre ochreous, peripheral lobes orange-brown, K+ purple. — **13. C. ehrenbergii**
6. Apothecia black, usually pruinose. — **10. C. aegyptiaca** var. **circinans**
- Apothecia orange, reddish to brown, naked — 7
7. On siliceous rocks. Thallus yellowish green or greyish green, K+ yellow. — **12. C. carphinea**
- On calcareous rocks. Thallus not as above, K– — 8
8. Thallus starchy white. Apothecia reddish brown. — **15. C. erythrocarpa**
- Thallus sand-coloured. Apothecia orange. — **14. C. erythrina** var. **pulvinata**

10. Caloplaca aegyptiaca (Müll. Arg.) Stein. var. **circinans** Stein., Ann. Nat. Hist. Staatsmus. 34: 55 (1921). [Plate XIX: 5]

Thallus more or less orbicular, 1–2 cm across, usually several colonies intergrading and covering large areas, rimose-areolate; areoles rough verrucolose-granulose; peripheral zone more or less lobate; verrucoles whitish grey and sand-coloured in between, variable in thickness (160–780 μ). Upper cortex paraplectenchymatous, 0–160 μ thick, covered with a thick, greyish granular amorphous layer. Lower cortex absent.

Apothecia at first partly immersed, roundish and even with the thallus, finally broadly adnate and deformed by compression, up to 1 mm across. Disc black, usually pruinose, concave or plane. Thalline margin whitish, persistent, 90–100 μ thick. Exciple 10–30 μ at the base, 70–110 μ above, colourless except for the fuliginose external cells. Hymenium *c.* 100 μ thick. Epithecium gelatinous-fuliginose. Hypothecium colourless, *c.* 300 μ thick. Paraphyses septate, more or less uniform. Spores 8, colourless, placodiomorph, 10.5–16.5 × 7.5 μ; isthmus 1.5–3 μ.

Reactions: Cortex usually K–, but K+ purple in certain spots, C–, Pd–; epithecium K+ violet.

Habitat: On limestone and dolomite. C. and W. Negev.

Distribution: Irano-Turanian and Saharo-Arabian.

11. Caloplaca aurantia (Pers.) Hellb. var. **aurantia** Poelt, Mitt. Bot. Staatssamml. München 11 : 22 (1954). [Plate XX : 1]

Thallus crustose, orbicular, up to 12 cm across, orange or yellowish orange, firmly attached to the substrate; centre rimose-areolate; areoles up to 1 mm across, plane or slightly convex; peripheral 2–3 mm lobate; lobes contiguous, approaching, plane or convex with obtuse or rounded tips; surface densely to moderately pruinose except for the lobate zone. Upper cortex colourless, paraplectenchymatous, thin to almost absent or up to 65 μ thick, usually covered with a granular, orange-brown layer. Lower cortex absent.

Apothecia numerous and usually crowded, centrally located, sessile, up to 1.5 mm across. Disc darker than the thallus, at first plane, finally convex. Thalline margin concolourous with the thallus, at first prominent, later disappearing. Exciple somewhat darker, thin, persistent. Hymenium and hypothecium colourless, 80–160 μ thick. Epithecium crystalloid, orange-yellow. Paraphyses septate; cells gradually shorter and thicker towards the apex. Spores 8, colourless, placodiomorph, lemon-shaped to rhomboid (Fig. 17 B), 10–12 × 3–7.5 μ.

Reactions: Thallus and apothecia K+ purple; contains parietin (Pl. XXVIII : 3).

Habitat: Very common and abundant on calcareous rocks. Upper and Lower Galilee, Mt. Carmel, Mt. Gilboa, Coast of Carmel, Shefela, Judean Mountains, N., C. and W. Negev.

Distribution: Mediterranean; also found in the dry warm regions of the Medio-European territory.

12. Caloplaca carphinea (Fr.) Jatta, Syll. Lich. Ital. 241 (1900). *Parmelia carphinea* Fr., Lichgr. Europ. Reform. 110 (1831). [Plate XIX : 3]

Thallus crustose, yellowish green or greyish green; centre more or less regularly areolate; areoles 0.2–0.5 (–0.7) mm across, plane or nearly so, black-rimmed; periphery of contiguous lobes, 1.5–2.5 mm long, 0.2–0.5 mm wide, with obtuse ends.

Apothecia abundant, up to 0.5 mm across, lecanorine, at first immersed, later adnate. Disc reddish brown, naked, at first plane, later convex with the thalline margin almost excluded. Hymenium colourless, 40–50 μ thick. Epithecium granulose, yellowish brown. Hypothecium colourless, 60–90 μ thick. Spores 8, colourless, placodiomorph, ovoid, 8–12 × 4–7 μ.

Reactions: Thallus K+ yellow, C+ yellow; medulla C+ yellow; contains gyrophoric acid (Pl. XXV : 5), atranorin (Pl. XXV : 1) and traces of usnic acid; apothecia K+ purple.

Habitat: On siliceous rocks. Samaria.

Distribution: Mediterranean.

13. Caloplaca ehrenbergii (Müll. Arg.) Zahlbr., Cat. Lich. Univ. 7 : 231 (1931). *Amphiloma ehrenbergii* Müll. Arg., Rev. Myc. 2 : 41 (1880). [Plate XX : 3]

Thallus crustose, up to 20 cm across or more, 0.5–2 mm thick; centre areolate; areoles 0.5–1 mm across, sometimes almost regularly quadrangular; periphery lobate; lobes up to 0.5 cm long and 1 mm broad, contiguous, more or less uniform, finger-shaped or furcate; centre ochreous; lobes orange-brown, yellowish green when growing in the shade. Cortex 15–25 μ thick, paraplectenchymatous, colourless but for the uppermost zone which is encrusted with orange-brown crystals. Lower cortex lacking.

Apothecia up to 2.5 mm across, sessile, numerous, finally occupying almost all the areoles except those near the lobate region. Disc plane, brown, epruinose. Thalline margin concolourous with the centre of the thallus, entire, persistent, elevated, 120–150 μ thick. Hymenium colourless. Epithecium dark brown, crystalloid. Hypothecium colourless, 130–160 μ thick. Paraphyses septate, single or double branched above. Spores 8, colourless, placodiomorph, ellipsoid or sometimes somewhat spherical, 7.5–10 × 3–6 μ.

Reactions: Thallus K+ purple, C+ purple; apothecia K+ purple.

Habitat: Common on flint. W. and C. Negev.

Distribution: Reported only from the deserts of Egypt.

14. **Caloplaca erythrina** (Müll. Arg.) Zahlbr. var. **pulvinata** (Müll. Arg.) Zahlbr., Cat. Lich. Univ. 7: 238 (1931). *Amphiloma erythrinum* var. *pulvinatum* Müll. Arg., Rev. Myc. 2: 42 (1880).

Thallus crustose, more or less orbicular, up to 2 cm across, solitary or several fused together, sand-coloured; centre areolate, mostly occupied by apothecia; circumference radiate-lobate; lobes 1–1.5 mm long, convex to plicate, contiguous; edges grey to dark grey. Cortex colourless, composed of densely intricate short-celled hyphae, 15–50 μ thick, covered with an amorphous layer *c.* 20 μ thick. Lower cortex lacking.

Apothecia numerous and crowded, slightly deformed by compression, up to 1 mm across. Disc poriform in young apothecia and discoid-plane when mature, orange, epruinose. Thalline margin persistent, entire, concolourous with the thallus. Hymenium colourless. Epithecium granular, yellow. Exciple colourless, composed of radiate hyphae *c.* 12 μ at the base, up to 35 μ and with yellow inspersed granules above. Hypothecium colourless. Spores 8, placodiomorph, colourless, 12 × 5 μ; isthmus about 1/3 of spore length.

Reactions: Thallus K–; apothecia K+ violet.

Habitat: On calcareous rocks. C. Negev.

Distribution: Reported only from the Galata Desert in Egypt.

15. **Caloplaca erythrocarpa** (Pers.) Zw., Flora 45: 487 (1862). *Patellaria erythrocarpa* Pers., Ann. Wetter. Ges. 2: 12 (1801). *C. lallavei* (Clem.) Flag. [Plate XX: 4]

Thallus orbicular, 2–6 cm across, frequently several fusing, firmly attached to the substrate, rimose; surface granulose, starchy white, effuse or radiate-plicate towards the periphery, often with a blackish hypothallus. Cortex plectenchymatous, 30–50 μ thick, brownish; lower cortex absent.

Apothecia mainly in the centre, numerous to crowded, 0.8 mm across, innate to broadly adnate. Disc chestnut or reddish brown, epruinose, plane. Thalline margin paler than disc, at first somewhat elevated, persistent. Hymenium colourless, 70– 80 μ thick. Epithecium granular, yellowish. Hypothecium colourless, 60–80 μ thick. Paraphyses branched, clavate. Spores 8, colourless, ellipsoid, placodiomorph, 13–21 × 4–8 μ.

Reactions: Thallus K−; apothecia K+ violet.

Habitat: On dolomite. Upper Galilee, Judean Mountains.

Distribution: Mediterranean, extending into S. Medio-European territories; also reported from the British Isles.

16. Caloplaca murorum (Hoffm.) Th. Fr., Lichgr. Scand. 1 : 170 (1871). *Lichen murorum* Hoffm., Enum. Lich. 63 (1784).

Thallus rosette-shaped, 0.5–2.5 cm across; centre warty-areolate, usually entirely occupied by apothecia; periphery lobate; lobes plane or somewhat convex, simple or incised, adnate, contiguous, 0.3–2 mm long and 0.2–1.5 mm broad; upper surface yellow-orange or dark orange, smooth or slightly pruinose; lower surface white.

Apothecia numerous and crowded, central, sessile, 0.5–1 mm across. Disc at first plane, later convex, orange (usually darker than the thallus), epruinose. Thalline margin yellow-orange, entire, at first prominent, finally partly rolled back. Hymenium colourless, 50–90 μ thick. Epithecium granular, golden yellow. Hypothecium colourless, subtended by a continuous algal layer. Exciple colourless, paraphysoid, fan-shaped, 10–15 μ below, up to 30–50 μ above. Spores 8, placodiomorph, narrow to broadly ellipsoid, 11–16 × 3–8.5 μ; isthmus 1/5–1/3 of the spore length.

Reactions: Thallus and apothecia K+ purple; contains parietin (Pl. XXVIII : 3).

Habitat: On dolomite and soft calcareous rocks. Upper and Lower Galilee.

Distribution: Subcosmopolitan.

17. Caloplaca negevensis Reichert et Galun, Bull. Res. Counc. Israel 9 D : 138 (1960).

Thallus consisting of minute lemon-green squamules, solitary, dispersed or grouped, on a thin sand-coloured hypothallus; squamules 0.5–1 mm across, 310–350 μ thick; centre concave, greyish-pruinose; circumference lobate; lobes soon becoming sorediate. Cortex colourless, paraplectenchymatous, with algal cells inspersed and covered with an amorphous yellowish stratum 8–40 μ thick.

Apothecia rare and dispersed on the hypothallus, 250–300 μ across. Disc plane, orange. Thalline margin somewhat paler, disappearing or changing into small granules. Hymenium colourless, *c.* 60 μ thick. Hypothecium colourless, 30–40 μ thick. Spores 8, colourless, ellipsoid, placodiomorph, 12 × 7 μ.

Reactions: Thallus and apothecia K+ purple; contains parietin (Pl. XXVIII : 3).

Habitat: On pottery and dolomite. C. Negev.

18. Caloplaca conglomerata (Bagl.) Jatta, Syll. Lich. Ital. 255 (1900). *Callopisma conglomeratum* Bagl., Nuov. Giorn. Bot. Ital. 3 : 243 (1871). *C. squamulosa* (Wedd.) B. de Lesd.

Thallus blackish or brownish gray, composed of small imbricate squamulose lobes. Cortex paraplectenchymatous, 0–100 μ thick, covered by an amorphous greyish layer. Medulla cellular but less dense than the cortex; lower cortex absent.

Apothecia usually sparse, sessile, up to 1.5 mm across. Disc dark reddish brown, naked, plane or somewhat concave. Exciple black, thin, entire. Thalline margin concolourous with the thallus, prominent, entire, crenulate or incised, persistent. Hymenium 70–80 μ thick, colourless. Epithecium yellowish, granular. Hypothecium colourless. Paraphyses uniform, simple or furcate. Spores 8, colourless, placodiomorph, broadly ellipsoid and sometimes ovoid, 11.5–16 × 4.5–7.5 μ; isthmus 3–4.5(–6.5) μ.

Reactions: Thallus K–; apothecia K+ purple; contain parietin (Pl. XXVIII : 3); hymenial gelatine and asci I+ violet; spores I–; hypothecium I–.

Habitat: On basalt. Upper Jordan Valley.

Distribution: Mediterranean, extending into the Medio-European lowlands.

Section TRIOPHTHALMIDIUM Zahlbr. Thallus crustose. Cortex lacking. Spores becoming 4-celled.

19. Caloplaca ochracea (Schaer.) Flag., Mém. Soc. d'Emul Doubs 257 (1886). *Lecidea ochracea* Schaer., Naturw. 2 : 11 (1818). *C. tetrasticha* (Nyl.) Oliv. [Plate XX : 2]

Thallus subdeterminate, very thin to almost invisible, bright to ochreous yellow, felted, continuous or somewhat areolate, delimited by a black hypothallus.

Apothecia rather numerous, dispersed, sessile, up to 0.5 mm across, biatorine. Disc at first poriform, finally plane or concave. Exciple paler than the disc, thick, prominent, persistent. Hymenium colourless, 60–80 μ thick. Epithecium orange, granular. Hypothecium colourless or faintly yellowish, *c.* 65 μ thick. Paraphyses simple, capitate. Spores 8, colourless, ellipsoid, 10–13 × 5–7.5 μ; loculi in the form of a broad canal about 1/3 as wide as the spore, divided into 4 cells (Fig. 17 C).

Reactions: Thallus and apothecia K+ purple; contains parietin (Pl. XXVIII : 3).

Habitat: On calcareous rocks. Upper Galilee, Judean Mountains.

Distribution: Atlantic, Medio-European and Mediterranean.

3. FULGENSIA Mass. et De Not.

Thallus crustose, lobate-effigurate at the periphery, yellow to orange, K+ purple, usually corticated above and sometimes partly corticated below, adnate, without rhizines.

Apothecia sessile, with thalline margin and exciple. Disc reddish, K+ purple. Spores 8, colourless, simple or with one thin partition.

1. Thallus isidiate. **3. F. subbracteata**
– Thallus without isidia 2
2. Thallus yellow or orange-yellow. Spores ellipsoid. **1. F. fulgens**
– Thallus orange. Spores finger-shaped, sometimes with one broadened end. **2. F. fulgida**

1. Fulgensia fulgens (Sw.) Elenk., Lich. Flor. Ross. Med. 2 : 246 (1907). *Lichen fulgens* Sw., Nov. Act. Acad. Upsal. 4 : 246 (1784).

Thallus 1–2 cm across, mustard-yellow; centre crustose, rough and warty; periphery lobate-effigurate, adnate; lobes 1–2 mm long and 0.5–1.0 mm wide, contiguous or somewhat imbricate; tips roundish, entire or incised.

Apothecia rare, centrally located, up to 0.6 mm across, sessile. Disc reddish brown, epruinose, at first plane, later slightly convex. Thalline margin concolourous with the thallus, at first prominent, soon disappearing. Exciple concolourous with the disc or somewhat darker, entire or somewhat crenulate. Hymenium colourless, 60–80 μ thick. Epithecium yellowish brown, granular. Hypothecium colourless, 45–60 μ thick, subtended by dispersed groups of algal cells, laterally not reaching above the hypothecium. Paraphyses septate, branched; apices clavate. Spores 8, colourless, simple, ellipsoid, 9–16 × 4–5 μ.

Reactions: Thallus and apothecia K+ purple; contains parietin (Pl. XXVIII: 3).

Habitat: On loess. C. Negev.

Distribution: Mediterranean and Medio-European extending into Atlantic territories; also found in N. America.

2. **Fulgensia fulgida** (Nyl.) Szat., Degen., Fl. Velebitica 3: 372 (1938). *Placodium fulgidum* Nyl., Flora 48: 212 (1865). [Plate XXII: 1]

Thallus 2–5 cm across; centre crustose, verrucose-bullate, cracking with age into areoles; periphery lobate-effigurate; lobes 1–2 mm long and 0.5–2 mm wide, adnate, approaching, somewhat incised; tips roundish; centre orange, periphery yellow and slightly pruinose.

Apothecia numerous, mainly in the centre, sessile, 0.5–2 mm across. Disc dark orange-brown or dark brown, epruinose, at first plane, later convex, finally deformed. Exciple concolourous with the disc. Thalline margin concolourous with the thallus, at first prominent, later disappearing. Hymenium colourless, 40–60 μ thick. Epithecium yellowish brown, granular. Hypothecium colourless, 100–125 μ thick, subtended by an algal layer, laterally not reaching above the hypothecium. Paraphyses septate, branched; apices clavate. Spores 8, colourless, simple, finger-shaped, some with one end broadened, 9–18 × 4–5 μ.

Reactions: Thallus and apothecia K+ purple; contains parietin (Pl. XXVIII: 3).

Habitat: On mossy soil in fissures of calcareous rocks and spreading on to the rock. Upper Galilee, Coast of Carmel, Judean Mountains.

Distribution: Mediterranean.

3. **Fulgensia subbracteata** (Nyl.) Poelt, Mitt. Bot. Staatssamml. München 4: 400 (1962). *Placodium fulgidum* Nyl., Flora 48: 212 (1865). [Plate XXII: 2]

Thallus 1–3 cm across, more or less orbicular; centre crustose, densely covered by very small, more or less globose isidial outgrowths; periphery lobate; lobes 1–2 mm long, contiguous, adnate, somewhat convex, plicate with roundish tips; centre orange-yellow; periphery greenish yellow, epruinose, fine-granular.

Apothecia rare, sessile, up to 1 mm across. Disc dark orange, epruinose, plane. Exciple somewhat paler than the disc, persistent. Thalline margin concolourous with the thallus, evanescent. Hymenium and hypothecium as in *F. fulgida*. Spores simple, ellipsoid, 10–14 × 4–5 μ.

Reactions: Thallus and apothecia K+ purple; contains parietin (Pl. XXVIII: 3).

Habitat: On mossy soil and oolithic limestone. Lower Galilee, Judean Mountains.

Distribution: Mediterranean.

4. Teloschistes Norm.

Thallus fruticose, grey or yellow to orange and then K+ red, upright or decumbent; branches cylindrical or compressed, of radial or somewhat dorsiventral structure. Cortical hyphae longitudinal in direction. Rhizines absent. Apothecia lecanorine. Disc orange, K+ purple. Spores 8, colourless, placodiomorph.

1. **Teloschistes lacunosus** (Rupr.) Sav., Acta Inst. Bot. Acad. Sci. USSR, ser. 2, 2 : 313 (1935). *Ramalina lacunosa* Rupr., Mem. Acad. Sci. St. Petersbourg, ser. 6, 6 : 235 (1845). *T. brevior* f. *halophilus* (Elenk.) Oxn. [Plate XXI : 3]

Thallus fruticose, forming cushions 2–5 cm high; branches 2–15 mm broad, repeatedly and irregularly divided, obscure grey, tomentose, rigid, flattened or channeled with margins strongly bent down; lower surface usually wrinkled-reticulate. Upper and lower cortex composed of colourless hyphae parallel to the surface, or nearly so, and covered with a greyish brown amorphous stratum.

Often sterile, sometimes with many laminal and marginal sessile or substipitate apothecia, up to 4 mm across. Disc cup-shaped to deep-concave, orange, naked. Thalline margin concolourous with the thallus, tomentose, thick, elevated, persistent. Hymenium colourless, 60–70 μ thick. Epithecium orange, granular. Hypothecium faintly yellowish. Spores 8, colourless, placodiomorph, fusiform-ellipsoid, 10–12 × 4–6 μ.

Reactions: Thallus K–, contains no lichen acids; apothecia K+ purple, contain parietin (Pl. XXVIII : 3).

Habitat: On loess (loose or attached) and on dead *Zygophyllum* branches. C. and W. Negev.

Distribution: In steppe and desert areas of the Pontic territory; also reported from C. Spain.

5. Xanthoria Th. Fr.

Thallus foliose, horizontal or subascending, yellow to orange, K+ purple (contains parietin) dorsiventral; with rhizines or with the lower side of the lobes attached to substrate; corticated above and below. Cortex composed of hyphae perpendicular to the surface. Phycobiont *Trebouxia* (Fig. 1 A), located beneath the upper cortex. Apothecia lecanorine. Spores 8, colourless, placodiomorph.

1. Thallus greenish yellow, up to 3 cm across. Spore lumina drop-shaped (Fig. 17 E). **3. X. steineri**
– Thallus yellow to dark orange, usually more than 3 cm across. Spore lumina not drop-shaped (Fig. 17 D) 2
2. Thallus yellow to orange-yellow, soft. Isidia always absent. Lobes contiguous or imbricate. Apothecia always present. Very common. **1. X. parietina**
– Thallus dark orange to brownish orange, rather rigid. Isidia present or not. Lobes discrete, contiguous, entangled or overlapping. Apothecia rare or absent. Not very common. **2. X. aureola**

1. Xanthoria parietina (L.) Th. Fr., Lich. Arctoi, 67 (1860). *Lichen parietinus* L., Spec. Plant. 2 : 1143 (1753).

Thallus foliose, growing in rosettes or irregularly spreading, usually large; lobes 1–5 mm broad, horizontal, soft, appressed, approaching or imbricate; tips rounded, crenulate, sometimes slightly ascending; upper surface yellow to orange-yellow (greenish or greyish when growing in shade); lower side whitish, plicate; 100–200 μ thick. Upper cortex paraplectenchymatous, colourless, either covered with a crystalloid yellow layer or inspersed with yellow crystals. Algae, *Trebouxia decolorans* (Ahmadjian, 1960), in a continuous layer. Medulla composed of thick and more or less longitudinal hyphae. Lower cortex similar to the upper cortex, but without crystals.

Apothecia usually numerous, 1–5 mm across, sessile or somewhat elevated. Disc orange-yellow, concave to plane and flexuous, epruinose. Thalline margin somewhat paler than the thallus, rather thin, entire, at first prominent, finally excluded. Hymenium colourless, 80–90 μ. Epithecium bright yellow, granular. Hypothecium colourless or faintly yellow. Paraphyses septate, capitate. Spores 8, colourless, ellipsoid, placodiomorph (Fig. 17 D), 12–16 × 5–9 μ.

Reactions: Thallus and apothecia K+ purple; contains parietin (Pl. XXVIII : 3).

Habitat: On trees, rocks, roofs, walls and fences. Plentiful except in the Judean Desert and Negev.

Distribution : Cosmopolitan.

2. Xanthoria aureola (Ach.) Erichs., Verh. Bot. Ver. Prov. Brandenburg 72 : 39 (1930). *Parmelia aureola* Ach., Lichgr. Univ. 437 (1810).

Thallus foliose, growing in small or large rosettes, often several confluent, somewhat rigid, rather closely attached to the substrate by the lower surface of the lobes, without rhizines; lobes 200–400 μ thick, 0.5–5 mm long and 0.5–2 mm broad, discrete or somewhat entangled, adjacent or partly overlapping, without soredia, with or without isidia; upper surface dark orange or red to orange-brown; lower side whitish or pale orange. Upper and lower cortex paraplectenchymatous; upper cortex inspersed or covered with orange-yellow crystals. Medullary hyphae more or less parallel to the surface.

Apothecia up to 2 mm across, adnate. Disc dark orange, at first somewhat concave and marginate, later plane and somewhat immarginate. Hymenium and hypothecium colourless. Epithecium of orange crystals. Spores 8, colourless, placodiomorph, ellipsoid or subglobose.

Reactions : Thallus and apothecia K+ purple; contains parietin (Pl. XXVIII : 3).

Var. **aureola** Poelt, Mitt. Bot. Staatssamml. München 4 : 568 (1962) [Plate XXI : 2]. Thallus brownish orange, centre composed of small knotty warts. Apothecia absent or very rare.

Habitat : Common and abundant on basalt. Upper Galilee.

Distribution : Mediterranean, Medio-European and Atlantic.

Var. **ectaniza** (Nyl.) Poelt, Mitt. Bot. Staatssamml. München 4 : 568 (1962). *Lecanora elegans* var. *ectaniza* Nyl., Flora 66 : 105 (1883) [Plate XXI : 1]. Thallus dark orange, lobes 0.5–2 mm long and 0.5–0.6 mm broad, discrete or somewhat entangled, bullate-crenulate at the centre. Apothecia not seen.

Habitat: On calcareous rocks and mossy soil. Upper Galilee.

Distribution: Mediterranean and Irano-Turanian.

Var. **isidioidea** Beltr., Lich. Bassan. 103 (1858) [Plate XXI: 4]. Centre entirely covered with dark orange granuliform isidia; lobes somewhat paler, 2–5 mm long and 1–2 mm broad, convex, contiguous to partly overlapping; margins flat and attached to the substrate or partly free.

Apothecia very rare in the Negev, more common in the Upper Galilee. Spores subglobose, 9–10.5 × 6 μ.

Habitat: On calcareous rocks. Upper Galilee, Mt. Gilboa, Coast of Carmel, N. and C. Negev.

Distribution: Mediterranean (in Israel also in Irano-Turanian and Saharo-Arabian territory).

3. **Xanthoria steineri** Lamb, Jour. Bot. 74: 350 (1936).

Thallus foliose, pulvinate-orbicular, 1–3 cm across, 90–150 μ thick, usually several confluent (especially on the northern exposed side), greenish yellow; lobes crenate, shallowly subdivided with rounded, slightly raised, orange-coloured margins; lower side pale, rather loosely attached to the substrate. Upper cortex paraplectenchymatous, colourless and clear, covered with a yellowish granular stratum. Algal layer continuous, more or less uniform. Medullary hyphae thick-walled, parallel to the surface or nearly so. Lower cortex similar to the upper, somewhat thinner.

Apothecia abundant and crowded in the centre, up to 2 mm across, constricted at the base. Disc orange, concave to plane. Thalline margin entire, prominent, more or less persistent, concolourous with the thallus. Hymenium 50–75 μ thick, colourless except for an orange-yellow epithecium. Hypothecium colourless, composed of thick-walled hyphae more or less parallel to the surface. Paraphyses septate, furcate; apical cells enlarged. Spores 8, colourless, placodiomorph, ellipsoid, with the lumina in the form of two drops connected at their acute ends (Fig. 17 E), 10–13.5 × 5–7 μ.

Reactions: Thallus and apothecia K+ purple; hymenium I+ blue; hypothecium I–.

Habitat: On branches of *Lycium* sp. C. Negev.

Distribution: Reported from the Bahrein Island (Persian Gulf).

VERRUCARIACEAE*

Thallus crustose, superficial or growing within the upper layers of the substrate. Algae belonging to the Chlorophyceae. Perithecia solitary with an apical pore for spore ejection. Paraphyses usually gelatinizing or lacking from the mature perithecium.

* According to Zahlbruckner (1926).

VERRUCARIA (Wigg.) Th. Fr.

Thallus crustose, growing within or upon the substrate, stratified. Phycobiont *Myrmecia* (Fig. 1 B), *Coccobotrys* or *Pleurococcus* (Fig. 1 C) (Ahmadjian, 1967). Perithecia solitary, entirely or half immersed or sessile, with a terminal pore, the ostiole. Exciple spherical, semispherical or flask-like, dark to colourless, often covered with a lid, the involucrellum, around the ostiole. Paraphyses soon gelatinizing. Spores 8, simple, colourless or rarely pigmented.

1.	Thallus partly endolithic, grey to rose-lilac.	**2. V. marmorea**
–	Thallus superficial, not coloured as above	2
2.	Thallus dark brown to blackish. Spores 10–18 μ long.	**1. V. fuscella**
–	Thallus olive-green to greyish. Spores 15–35 μ long.	**3. V. viridula**

1. Verrucaria fuscella Ach., Lichgr. Univ. 289 (1810). [Plate XXII : 3]

Thallus superficial, dark brown or blackish brown, small and somewhat orbicular or large and indeterminate, *c.* 1 mm thick, deeply cracked into areoles of more or less regular size; areoles plane, black-rimmed on account of the cup-shaped blackish brown medulla in which the cortex and algal layer are embedded (Fig. 18). Cortical and algal cells included in a colourless net-like layer 70–100 μ thick, covered by an amorphous yellowish brown layer 10–15 μ thick.

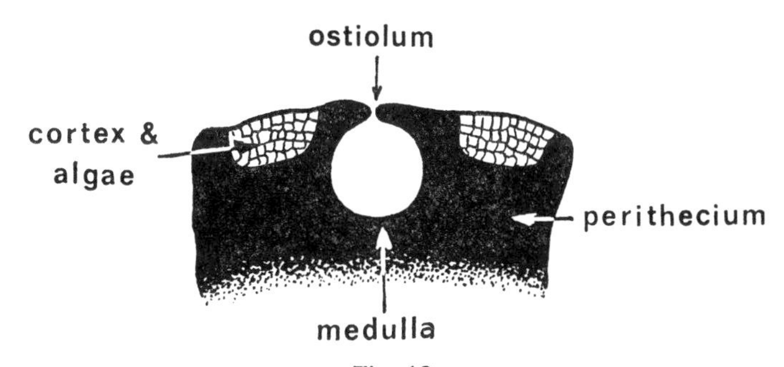

Fig. 18
Verrucaria fuscella
Section through thallus and perithecium (× 60)

Perithecia minute, immersed, scarcely visible, 1–3 in each areole. Ostiole black, level with the thallus, somewhat elevated or slightly depressed. Exciple brown below, pale near the ostiole. Spores 8, colourless, simple, ovoid-ellipsoid, 10–18 × 4–7.5 μ.

Reactions: Hymenium I+ red.

Habitat: Common on calcareous rocks. Upper Galilee, Mt. Carmel.

Distribution: Mediterranean, S. Medio-European; also found in the British Isles and N. America.

2. Verrucaria marmorea Ach., Flora 68 : 73 (1885). [Plate XXII : 4]

Thallus partly endolithic, superficial crust effuse, continuous, smooth, grey to rose-lilac, limited from other lichen specimens by a black line. Cortical layer red, 15–20 μ thick, underlain by groups of algal cells. Superficial part of the medulla composed of colourless, loosely intricate branched hyphae, 20–30 μ thick.

Perithecia numerous, black, at first somewhat emerging, finally immersed, leaving deep pits in the substrate. Exciple reddish black near the ostiolum, the rest pale. Spores 8, colourless, simple, ellipsoid or ovoid, 15–24 × 10–12.5 μ.

Reactions: Thallus K+ blue-green; hymenium I+ at first blue, rapidly turning reddish brown.

Habitat: On sun-exposed calcareous boulders. Upper and Lower Galilee, Mt. Gilboa.

Distribution: Mediterranean.

3. **Verrucaria viridula** Ach., Method. Lich. Suppl. 16 (1803).

Thallus superficial, rimose-areolate; areoles 0.2–0.5 mm across, polygonal, smooth or verrucolose, olive-green to grey, sometimes with a black hypothallus.

Perithecia numerous, black, *c.* 0.5 mm across and 0.6–0.7 mm deep, immersed. Exciple subglobose, black and thick above, thin below. Spores 8, simple, colourless, broadly ellipsoid, 15–35 × 8–15 μ; plasma appearing reticulate on account of the numerous vacuoles.

Reactions: Thallus K–; hymenium I+ violet.

Habitat: On oolithic limestone and calcareous rocks. Upper Galilee, Mt. Carmel.

Distribution: Temperate regions of the N. Hemisphere.

DERMATOCARPACEAE*

Thallus foliose, squamulose to subcrustose, umbilicate or rhizinate, corticate on one or both sides. Phycobiont belonging to the Chlorophyceae. Perithecia simple, more or less immersed in the thallus, opening by an apical pore for spore ejection.

DERMATOCARPON (Eschw.) Th. Fr.

Thallus foliose, squamulose or squamulose-areolate, umbilicate or with rhizines, corticate above or on both sides. Phycobiont *Hyalococcus* or *Myrmecia* (Fig. 1 B). Perithecia with a colourless or carbonaceous exciple. Paraphyses usually gelatinizing at maturity. Spores usually 8, rarely 16, colourless, simple.

1. Thallus foliose, umbilicate. **5. D. miniatum**
– Thallus squamulose or squamulose-areolate, attached by medullary or hypothalline hyphae 2
2. Thallus squamulose, brown to black, epruinose 3
– Thallus squamulose-areolate, greyish- or bluish-pruinose 5
3. Squamules up to 1 mm across, convex, black. **2. D. convexum**

* According to Zahlbruckner (1926).

- Squamules more than 1 mm across, plane or at least partly concave, pale to dark brown 4
4. Hymenium I–. **4. D. hepaticum**
- Hymenium I+ reddish brown. **3. D. desertorum**
5. Exciple carbonaceous. **6. D. monstrosum**
- Exciple colourless 6
6. On basalt. 2–2.5 mm thick. Hymenium I+ blue. **1. D. bucekii**
- On calcareous substrate. Up to 1 mm thick. Hymenium I–. **7. D. subcrustosum**

1. Dermatocarpon bucekii Nadv. et Serv., in Serv. Beih. Bot. Centralbl. 55 (b): 267 (1936). [Plate XXIII: 4]

Thallus an uneven crust, small or large, squamulose-areolated, attached to the substrate by thick blackish stipes; areoles 2–4 mm across and 2–2.5 mm thick, irregular, plane, convex or turgid, greyish-pruinose, some black-rimmed, separated by deep and wide fissures and subdivided by narrow, shallow, more or less incomplete cracks; lower side blackish; periphery partly lobate-effigurate; lobes plane, closely adnate, 1–1.5 mm broad and 3–4 mm long with somewhat crenate margins. Upper cortex colourless below, brown above, consisting of vertical cell rows. Algal cells densely aggregated beneath the cortex in vertical rows. Medulla consisting of loosely arranged roundish, colourless cells interpenetrated by aggregated and dispersed algal cells. Lower cortex and side walls of each areole dark brown.

Perithecia numerous in some areoles and rare in others, immersed, 200–250 μ across, spherical or somewhat elongated. Exciple colourless. Ostiole dark brown. Asci usually dissolved. Spores colourless, elongated or shaped like a falling drop, simple or 2-celled, with 1, 2 or many oil drops (all forms may appear in the same perithecium); 15–20 × 5–7 μ.

Reactions: Hymenium I+ blue; thallus negative to all reagents.

Habitat: On basalt, mainly in rock depressions and on rough surfaces. Upper and Lower Galilee, Upper Jordan Valley.

Distribution: Mediterranean, S. Medio-European and Pontic.

2. Dermatocarpon convexum Reichert et Galun, Bull. Res. Counc. Israel 9 D: 129 (1960).

Thallus squamulose; squamules 0.5–1 mm across, 150–500 μ thick, somewhat orbicular, convex, dark brown to blackish, contiguous; hypothallus black. Upper cortex composed of thin-walled vertical cell rows, colourless in the lower part, outermost 15–20 μ dark brown, partly covered with an amorphous yellowish brown layer. Algal layer continuous, distinctly limited upwards, indistinctly limited towards the medulla. Medulla cellular, translucent, with the lower side dark brown; lower cortex lacking.

Perithecia 1–2 per squamule, immersed with the apex somewhat emerging, globose, 250–400 μ across. Exciple pale to pale brown, dark brown near the ostiole. Asci numerous. Spores 8, colourless, simple, ellipsoid, with 1–2 oil drops, 10.5–13.5 × 6–7.5 μ.

Habitat: On calcareous rocks. C. Negev.

3. Dermatocarpon desertorum * Tom., Animadvers. System. ex Herb. Univ. Tomsk. 2 : 1 (1931).

Thallus squamulose; squamules 0.5–1.5 mm across; central squamules convex to tuberculate, plane or concave at the periphery; upper surface dark brown, smooth or cracked, epruinose; lower side pale except for a dark brown marginal zone, loosely attached to the substrate by a group of centrally located colourless hyphae 4–5 μ thick. Upper cortex composed of perpendicular cell rows, colourless except for an exterior brown zone, covered with a thin translucent amorphous layer. Algal layer continuous, vertically arranged. Medulla composed of loosely arranged roundish colourless cells delimited by a brownish obscure lower zone.

Perithecia 1–3 per squamule, immersed, subglobose. Exciple pale brown below, darker brown above. Spores 8, colourless, simple, subglobose, 15–17 × 10–12 μ.

Reactions : Hymenium I+ reddish brown.

Habitat : On thin layers of loess accumulating in fissures and depressions of Nubian sandstone. C. Negev.

Distribution : Reported once from Altai (China).

4. Dermatocarpon hepaticum (Ach.) Th. Fr., Nova Acta Reg. Soc. Sci. Upsal. 3(3): 355 (1861). *Endocarpon hepaticum* Ach., Kgl. Vetensk.-Akad. Nya Handl. 156 (1809). [Plate XXIII : 1]

Thallus squamulose; squamules 2–5 mm across and 140–300 μ thick, closely adnate, dispersed or several connate, at first roundish, later irregular-crenate, plane or somewhat concave; upper surface pale, dark or reddish brown, usually black-rimmed; lower side pale. Upper cortex composed of vertical cell rows, colourless except for the yellowish brown pigmented cell walls of the two uppermost cell rows, covered (especially in the Negev) with a thick, transparent, amorphous, stratum. Algae, *Myrmecia biatorelle* (Geitler, 1962), vertically arranged. Medulla cellular, colourless, with thick rhizoidal hyphae extending into the substrate.

Perithecia rare or frequent, immersed, 300–350 μ across. Exciple mostly colourless, brown near the dark ostiole. Spores 8, colourless, simple, ellipsoid, 12–16 × 5–7 μ.

Reactions : Hymenium I–; thallus negative to all reagents.

Habitat : On soil in crevices of rocks, common. Upper and Lower Galilee, Coast of Carmel, Judean Mountains, Judean Desert, N., C. and S. Negev, Arava Valley.

Distribution : Common in the temperate regions of the N. Hemisphere.

5. Dermatocarpon miniatum (L.) Mann, Bohem. Observ. Dispos. 66 (1825). *Lichen miniatus* L., Spec. Plant. 1149 (1753). [Plate XXIII : 2]

Thallus foliose, mono- or polyphyllous, umbilicate; leaves 1–7 cm across, shield-shaped, roundish or crenate-lobate, plane or with raised, or raised and revolute, margins;

* According to Weber (1962) *Dermatocarpon lachneum, D. hepaticum, D. rufescens* and *D. desertorum* should be assigned to a single species which varies in response to variations in soil texture and stability, shade, available moisture, and other changes in the habitat; he accepts the oldest name, **D. lachneum** (Ach.) A. L. Sm.

upper surface brownish and densely covered with a grey to dark grey pruina; lower side brown or blackish brown.

Perithecia numerous, immersed, globose, *c.* 200 μ across. Exciple colourless. Ostiole brownish, prominent. Paraphyses usually dissolved. Spores 8, simple, colourless, ellipsoid, 8–14 × 5–6 μ.

Reactions: Hymenium I+ blue; thallus negative to all reagents.

Habitat: Plentiful on basalt and rare on calcareous rocks, in shaded and damp situations. Upper Galilee.

Distribution: Subcosmopolitan.

6. **Dermatocarpon monstrosum** (Schaer.) Wain., in Természetr. Füzetek 22: 336 (1899). *Endocarpon miniatum* var. *monstrosum* Schaer., Lich. Helv. Spic. 7: 349 (1836). [Plate XXIII: 3]

Thallus squamulose-areolate; squamules dispersed or most often contiguous and separated by deep fissures, 3–6 mm across and up to 2 mm thick, polygonal, plane to somewhat undulate; upper surface brownish and bluish-pruinose; lower side black, closely attached to the substrate. Cortical and algal cells forming one layer covered by an amorphous, brownish stratum. Medulla colourless, cellular. Lower cortex dark brown.

Perithecia rare, immersed, *c.* 200 μ across, globose or nearly so. Exciple carbonaceous. Spores 8, elongate-ellipsoid, simple, colourless, 22–28 × 6–8.5 μ.

Reactions: Hymenium I+ blue; thallus negative to all reagents.

Habitat: Rather common on dolomite and limestone. Upper Galilee.

Distribution: Mediterranean and S. Medio-European.

7. **Dermatocarpon subcrustosum** (Nyl.) Zahlbr., Cat. Lich. Univ. 1: 236 (1922). *Endocarpon cinerascens* var. *subcrustosum* Nyl., Memoir. Soc. Sci. Nat. Cherbourg 2: 340 (1854). [Plate XXIII: 5]

Thallus squamulose-areolate; areoles 2–3 (–4) mm across, 0.3–1.0 mm thick, irregular, more or less plane, greyish-pruinose, black-rimmed, separated by broad and deep fissures and subdivided by narrow, sometimes incomplete, incisions; periphery with occasional irregularly incised lobes. Upper cortex colourless, composed of thin-walled perpendicular cell rows covered with an amorphous layer. Algal layer more or less uniform, also in perpendicular rows. Medulla colourless, cellular. Lower cortex and side walls of each areole blackish brown.

Perithecia numerous in each areole, immersed, subglobose, 180–230 μ across. Exciple colourless. Ostiole brown. Asci usually dissolved. Spores simple, colourless, ellipsoid, 13–15.5 × 6.5–7.5 μ.

Reactions: Hymenium I–; thallus negative to all reagents.

Habitat: On calcareous rocks. Upper Galilee.

Distribution: Mediterranean, S. Medio-European and Mesopotamian.

CYPHELIACEAE

Thallus crustose, uniform; algae belonging to the Chlorophyceae. Apothecia sessile with exciple and thalline margin, or with thalline margin only. Hymenial region breaking up to form an adnate mazaedium when mature.

Carlosia Samp.

Thallus crustose, heteromerous. Cortex composed of conglutinate, perpendicular hyphae. Apothecia lecanorine, sessile, mazaedium blackish. Asci with 8 spores. Spores simple, at first colourless, later dark, spherical.

Only one species known.

1. **Carlosia lusitanica** Samp., Nota apres. ao Congr. de Salamanca, Pôrto 1 (1923). [Plate XXII : 5]

Thallus crustose, indeterminate, 2–4 cm across, ash-grey, rimose-areolate, 0.5–1.0 mm thick; surface dull and rough. Cortex colourless, irregular in width, composed of nearly perpendicular hyphae with a more or less uniform layer of fine greyish crystals in the centre; outermost part amorphous-necrotic. Algal layer contiguous, more or less uniform. Medulla colourless, compactly intricate, predominantly perpendicular, with crystals and substrate particles.

Apothecia abundant, orbicular, up to 1 mm across and about 0.5 mm deep, broadly adnate, usually occupying an entire areole, trapezoid in section. Thalline margin somewhat darker than the thallus, smooth, slightly glossy, except for the whitish, rough rim that surrounds the disc. Disc black, moderately elevated above the margin, plane or convex, pulverulent. Hypothecium clear, colourless to faintly yellowish below. Hymenium covered by a mass of free spores (mazaedium). Paraphyses colourless, septate, branched. Spores greenish brown to brownish black, spherical, 12.5–15.5 μ across; spore wall sculptured with tiny projections.

Reactions: Thallus K–, C–, KC+ red; contains collatolic acid (?) (Pl. XXVII : 2).

Habitat: On basalt. Rare. Esdraelon Plain (Giv'at ha-Moreh).

Distribution: *C. lusitanica* is the only species of the genus *Carlosia* and has been reported only from Lusitania (Portugal).

ARTHONIACEAE

Thallus crustose, undifferentiated, often hypophloeodal or lacking. Phycobiont *Trentepohlia* (Fig. 1 E). Apothecia roundish or elongate or irregular, immarginate. Spores 8, colourless, 2- to multiseptate or muriform. Paraphyses branched and intricate.

ARTHONIA Ach.

Thallus crustose, hypo- or epiphloeodal. Phycobiont *Trentepohlia*. Apothecia innate, immarginate, roundish, elongate or irregular. Spores 2- to multiseptate, colourless.

Arthonia melanophthalma Duf., Memoir. Soc. Sci. Nat. Cherbourg 2 : 336 (1854). [Fig. 19]

Thallus partly hypophloeodal and partly epiphloeodal, indeterminate, whitish, irregularly rimose-areolate; areoles 0.2–1 mm across; firmly attached to the substrate when on intact branches, scaling off on loose bark; most of the algal cells and medullary hyphae penetrating the peridermal layers of the substrate and separating them.

Apothecia numerous to crowded, 0.3–0.8 mm across, roundish, polygonal or elongate. Disc black, naked, at first plane and level with the thallus, later convex and eventually emerging. Hymenium faintly yellowish. Epithecium blackish, covered with a colourless, amorphous layer. Hypothecium in contact with the substrate, black or dark brown. Paraphyses very thin, branched and densely intricate. Spores 8, colourless, 3–4-septate, drop-shaped, 13–15 × 4–5 μ; the uppermost cell largest.

Reactions: Thallus K–, C–, Pd–; hymenial gelatine I+ greenish blue; asci and spores I–.

Habitat: On *Ceratonia siliqua*. Mt. Carmel.

Distribution: Mediterranean.

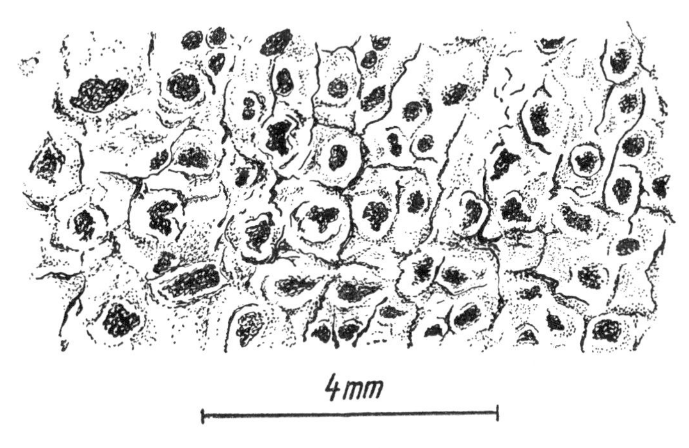

Fig. 19
Arthonia melanophthalma

ARTHOPYRENIACEAE

Thallus crustose, poorly differentiated, superficial or innate. Phycobiont *Trentepohlia* (Fig. 1 E). Fruiting bodies perithecium-like, solitary or clustered. Paraphyses anastomose, persistent or gelatinizing.

Arthopyrenia Mass.

Thallus crustose, poorly developed, homoeomerous, superficial or growing partly within the substrate. Phycobiont *Trentepohlia*. Fruiting bodies solitary, prominent or immersed, spherical or semispherical. Exciple carbonaceous. Ostiole apical. Paraphyses anastomose, persistent or dissolving. Spores 8, colourless, oblong to oval, transversely 1–5-septate.

1. Arthopyrenia conoidea (E. Fr.) Zahlbr., Engler-Prantl, Nat. Pfl.-Fam. 1(1): 65 (1905). *Verrucaria conoidea* E. Fr., Lich. Eur. Ref. 435 (1831).

Thallus greyish or whitish, tartareous, very thin, scarcely visible.

Fruiting bodies numerous, black, almost globose, *c.* 1 mm across, prominent, somewhat immersed at the base or almost entirely sessile. Exciple 75–85 μ thick, brownish black, surrounding the colourless hymenium except at the base. Paraphyses distinct, colourless, aseptate, slender, anastomose. Asci cylindrical. Spores 8, uniseriate, colourless, oblong or broadly ellipsoid, 1-septate; septum somewhat thickened, 20–25.5 × 7.5–10.2 μ.

Reactions: Thallus K–, C–; hymenium I–.

Habitat: On oolithic limestone. Mt. Carmel.

Distribution: Medio-European and Mediterranean.

OPEGRAPHACEAE

Thallus crustose, homoeo- or heteromerous. Phycobiont *Trentepohlia* (Fig. 1 E). Apothecia roundish or variously shaped. Exciple well developed. Spores 8, 2–16-celled, septae thin, cells cylindrical.

Opegrapha Humb.

Thallus crustose, uncorticated. Phycobiont *Trentepohlia*. Apothecia sessile, adnate or innate, roundish or elongate (lirelliform). Exciple carbonaceous. Hypothecium dark or colourless. Paraphyses branched and intricate. Spores 8, colourless, 2- to multicellular; cells cylindrical.

1. Corticolous. Thallus mainly hypophloeodal. — **1. Opegrapha atra**
– Saxicolous. Thallus superficial. — **2. Opegrapha grumulosa**

1. Opegrapha atra Pers., in Usteri, Neue Ann. d. Bot. 1: 30 (1794).

Thallus existing as small roundish whitish grey patches, mainly hypophloeodal.

Apothecia numerous, adnate to partly innate, very variable in shape and size, up to 3 mm long and 0.2–0.3 mm wide, simple or branched, linear, flexuous or polygonal. Disc slit-like, narrow, black, epruinose. Exciple thick, black, elevated, wavy. Hypothecium dark reddish brown, intergrading with the exciple. Hymenium colourless, 60–80 μ thick. Epithecium brownish, gelatinous. Spores 8, colourless, obovate-fusiform, straight, 3-septate, 12–18 × 4 μ.

Reactions: Thallus K–; hymenial gelatine I+ blue; asci I+ reddish.
Habitat: On branches of fig trees. Rare. Sharon Plain.
Distribution: Common in Medio-European and Atlantic territories and rare in the Mediterranean region; also found in N. America.

2. **Opegrapha grumulosa** Duf., Journ. Phys., Chim. d'Hist. Nat. 87 : 214 (1818).
Thallus indeterminate, brittle, white, farinose, cracked in the centre, sometimes with a very thin black hypothallus.
Apothecia abundant, crowded, at first immersed and roundish, then sessile, elongate or angular or deformed. Disc black, pruinose. Exciple black, level with the disc or somewhat above it, epruinose or less pruinose than the disc, flexuous, crenulate or incised, persistent. Hymenium yellowish, 85–90 μ thick. Epithecium olive-green, gelatinous. Hypothecium brownish black, 250 μ thick. Paraphyses conglutinate, branched above, with olive-green apical cells. Spores 8, colourless, oblong-fusiform, 4-celled, sometimes 3- or 5-celled, 13.5–16 × 3 μ; cells cylindrical.
Reactions: Thallus K–, C+ red; hymenium I+ red.
Habitat: On oolithic limestone. Mt. Carmel.
Distribution: Mediterranean, extending into Atlantic and S. Medio-European territories.

DIRINACEAE*

Thallus crustose, stratified and corticated. Phycobiont *Trentepohlia* (Fig. 1 E). Apothecia with exciple and thalline margin. Hypothecium black. Spores 8, colourless or brown, spindle-shaped, 4–8-celled.

Dirina E. Fr.

Thallus crustose, attached to the substrate by medullary hyphae. Cortex composed of aseptate hyphae perpendicular to the surface. Phycobiont *Trentepohlia*. Apothecia with a thin exciple and a thick thalline margin. Hypothecium carbonaceous. Paraphyses unbranched. Spores 8, colourless, 4–8-celled.

1. **Dirina ceratoniae** (Ach.) E. Fr., Lichgr. Europ. Reform. 194 (1831). *Lecanora ceratoniae* Ach., Lichgr. Univ. 361 (1810). [Plate XXII : 6]
Thallus crustose in roundish patches composed of irregular tuberculate areoles, greyish, becoming creamy white with time, variable in thickness (60–450 μ). Cortex composed of aseptate, more or less perpendicular, colourless hyphae 20–40 μ thick, inspersed with crystals and covered with a thick crystalline layer. Algal layer continuous, 25–200 μ thick. Medulla loose, obscure.
Apothecia numerous to densely crowded, adnate, roundish, 0.5–2 mm across. Disc black with a thick, farinose, white cover which sometimes scales off. Thalline margin

* According to Zahlbruckner (1926).

thick, entire, concolourous with the thallus. Hymenium colourless, 100–130 μ thick. Epithecium dark, 40–60 μ thick. Hypothecium black. Exciple thin, brownish black above, colourless at the base. Paraphyses simple, unbranched. Spores 8, colourless, fusiform, 4-celled, 20–29 × 3.5–5.5 μ.

Reactions: Thallus K± yellowish green, C+ orange-red, Pd–; contains erythrin (Pl. XXIV: 3); hymenium I+ blue.

Habitat: Abundant on *Ceratonia siliqua.* Mt. Carmel.

Distribution: Mediterranean.

GLOSSARY

Acicular Slender, needle-like.

Apothecium An open disc- or cup-shaped fruiting body.

Areole A small polygonal or roundish division of a surface, separated from others by narrow cracks. A thallus composed of areoles is areolate.

Areolate See Areole.

Ascus The sac-like structure which contains the ascospores and is located in the hymenial region of the apothecium.

Biatorine Referring to the apothecium: with a colourless or bright-coloured exciple and without a thalline margin.

Capitate With a head-like tip.

Cilia Marginal hair-like structures.

Clavate Club-shaped.

Contiguous Touching and continuous.

Cortex The outer layer of the thallus.

Corticate With a cortex.

Corticolous Growing on bark.

Crenate, Crenulate With small rounded projections along the edge.

Crustose Crust-like, closely adpressed to the substrate.

Determinate Having a definite outline.

Dichotomous Branching into two equal branches.

Dorsiventral Having distinct upper and lower sides.

Ecorticate Lacking a cortex.

Esorediate Lacking soredia.

Effuse Referring to the thallus: indeterminate and spreading.

Effigurate Referring to the thallus: orbicular and adpressed, lobed at the circumference.

Endolithic Growing immersed in stone.

Entire Margin not crenate or otherwise divided.

Epithecium The layer covering the hymenium.

Epiphloeodal Growing on the surface of bark or wood.

Euparaplectenchymatous Referring to a type of exciple in the Collemataceae: consisting of strongly conglutinate, somewhat isodiametric, globose to usually polygonal cells (Fig. 2 C).

Euthyplectenchymatous Referring to a type of exciple in the Collemataceae: consisting of densely packed, not conglutinate, filiform hyphae with somewhat elongated cells (Fig. 2 A).

Exciple The margin around the apothecium which is of purely fungal structure without any associated algae; also known as the proper margin.

Farinose Very finely powdery, flour-like.

Foliose Referring to the thallus: leaf-like.

Fruticose Referring to the thallus: bushy or shrubby, attached by a single basal point to the substratum.

Fusiform Spindle-shaped.

Hammada Desert with stony or gravelly cover.

Heteromerous With the fungal and algal components in definite stratified zones.

Homoeomerous With the fungal and algal components lacking any definite order.

Hymenium The layer of the apothecium which is composed of the asci and the paraphyses.

Hypha A fungal filament.

Hypothecium A layer in the apothecium which lies directly below the hymenium.

Hypophloeodal Growing inside the bark.

Hypothallus An undifferentiated growth of hyphae below and / or around the lichen thallus.

Involucrellum The carbonized thalline margin of the perithecium in some pyrenolichens.

Isidium A corticated outgrowth from

the thallus, which apparently functions in the asexual reproduction of lichens.

Isthmus A canal connecting the two lumens of a placodiomorph spore.

Labriform Referring to the soralia: consisting of lip-shaped patches at the lobe tips.

Laciniate Cut into narrow lobes.

Lecanorine Referring to the apothecium: having a thalline margin.

Lecideine Referring to the apothecium: having a dark-coloured exciple and lacking a thalline margin.

Laminal Located on the surface of the thallus.

Lobate Divided into lobes.

Lobules Small lobes.

Marginal Located on the edge or margin.

Mazaedium A powdery mass of spores freed from the asci on the surface of the hymenium.

Medulla The loose hyphal layer in the interior of the thallus.

Mischoblastiomorph Referring to spores: with two funnel-shaped lumina and a thickened cross-wall, with or without a partition wall (Fig. 13 A).

Muriform Multicellular spores with transverse and longitudinal partitions like the masonry of a wall.

Orbicular Circular, round.

Ostiole A small opening in the perithecium through which the spores escape.

Paraphysis (pl. **paraphyses**) Sterile hypha among the asci in the hymenial layer of the apothecium.

Paraplectenchymatous Referring to the hyphose layer: cellular, the cells having isodiametric lumina.

Perithecium An immersed roundish or flask-shaped fruiting body with a small opening.

Phycobiont The algal component in the lichen thallus.

Placodiomorph Referring to spores: having two lumina separated by a thick cross-wall and interconnected by a narrow canal (isthmus), without any visible partition wall (Fig. 17 D).

Plectenchymatous Referring to the hyphose layer: not differentiated into a cellular tissue.

Plicate Folded, somewhat as in a fan.

Podetium The erect stalk of the Cladoniae.

Polarilocular Referring to 2-celled spores: having a thickened cross-wall and the lumina located at the two opposite ends.

Proper margin see Exciple.

Prosoplectenchymatous Referring to the hyphose layer: cellular, the cells having longish lumina.

Pruina A fine powdery covering.

Pruinose Covered with a pruina.

Pseudocortex A cortex-like outer layer in which the hyphae are distinct but not differentiated into prosoplectenchymatous or paraplectenchymatous tissue.

Pseudocyphellae Breaks in the cortex through which medullary hyphae protrude and appear as white dots on the surface of the thallus.

Pyrenocarpous Lichens having perithecia as fruiting bodies.

Rimose Having a cracked surface.

Rhizine A root-like hyphal strand extending from the lower cortex and anchoring the thallus to the substrate.

Rosette A circular arrangement of lobes.

Saxicolous Growing on rocks.

Sessile Referring to apothecia: lacking a stalk and arising directly from the thallus.

Simple Not divided or branched.

Soralium A delimited mass of erupted soredia.

Soredium A minute powdery or granulose structure containing a group of algae surrounded by fungal hyphae but lacking a cortex; apparently functions in the asexual reproduction of lichens.

Squamule Small thalline lobe.

Stipitate Having a stalk.

Stroma type Referring to the apothecium: consisting of several fruiting bodies opening by a small pore.

Subparaplectenchymatous Referring to a type of exciple in the Collemataceae: consisting of somewhat conglutinate, swollen, oval or oblong, sometimes globose cells (Fig. 2 B).

Terricolous Growing on soil.

Thalline margin The margin around the apothecium which is formed by the thallus and which contains algal cells.

Thallus The vegetative part of the lichen.

Umbilicate Referring to the thallus: having a navel-like depression near the centre, opposite the single hold-fast.

Urceolate Urn-shaped.

Verruca A granular wart-like part of the thallus.

BIBLIOGRAPHY

Ahmadjian V. (1960) 'Some New and Interesting Species of *Trebouxia,* a Genus of Lichenized Algae', *Am. J. Bot.,* 47 : 677–683.

— (1967) *The Lichen Symbiosis,* Blaisdell, New York.

Asahina Y. (1936–1940) 'Mikrochemischer Nachweis der Flechtenstoffe', I–IX, *Jap. J. Bot.,* 12–16.

Asahina Y. & S. Shibata (1954) *Chemistry of Lichen Substances,* Japan Society for the Promotion of Science, Tokyo.

Degelius G. (1954) 'The Lichen Genus *Collema* in Europe', *Symb. Bot. Upsal.,* 13 : 2.

Dughi R. (1952) 'Un problème de lichénologie non résolu — L'origine et la signification de l'apothécie lécanorine', *Ann. Fac. Sci. Marseille,* Ser. 2, 21 : 219–243.

Ericksen C. F. E. (1957) *Flechtenflora von Nordwestdeutschland,* Fischer Verlag, Stuttgart, p. 411.

Galun M. (1966) 'Lichens of the Galilee and their Chemical Constituents', *Israel J. Bot.,* 15 : 58–63.

Galun M. (1967a) 'A New Species of *Catillaria* from Israel', *The Lichenologist,* 3 : 423–424.

— (1967b) 'A New Location for *Gonohymenia mesopotamica* J. Stein', *The Bryologist,* 70 : 330–332.

Galun M. & H. Lavee (1966) 'Lichens from Har Meron (Jebel Jermak), Upper Galilee', *The Bryologist,* 69 : 324–333.

Geitler L. (1934) 'Beiträge zur Kenntnis der Flechtensymbiose', IV–V, *Arch. Protistenk.,* 82 : 51–85.

— (1960) 'Über Flechtenalgen', *Schweiz. Z. Hydrol.,* 22 : 130–135.

— (1962) 'Über die Flechtenalge *Myrmecia biatorella', Öst. Bot. Z.,* 109 : 41–44.

Hakulinen R. (1954) 'Die Flechtengattung *Candelariella* Müller Argovensis', *Ann. Bot. Soc. Zool.-Bot. Fenn. Vanamo,* 27(3) : 1–127.

Hale M. E. (1964) 'The *Parmelia conspersa* Group in North America and Europe', *The Bryologist,* 67 : 462–473.

— (1967) *The Biology of Lichens,* Arnold, London.

Henssen A. (1963) 'The North American Species of *Massalongia* and Generic Relationships', *Can. J. Bot.,* 41 : 1331–1346.

Lettau G. (1937) 'Monographische Bearbeitung einiger Flechtenfamilien', *Beih. Repert. Spec. Nov. Regni Veg.,* 69 : 1–250.

Luttrell E. S. (1955) 'The Ascostromatic Ascomycetes', *Mycologia,* 47 : 511–532.

Maas Geesteranus R. A. (1952) 'Revision of the Lichens of the Netherlands, II: Physciaceae', *Blumea,* 7 : 206–287.

Müller Argovensis J. (1884) 'Lichens de Palestine', *Rev. Mycol.,* 6 : 12–15.

Nannfeldt J. A. (1932) 'Studien über die Morphologie und Systematik der nichtlicheniesierten inoperculaten Discomyceten', *Nova Acta R. Soc. Scient. Upsal.,* Ser. 4, 8 : 1–368.

Nylander W. (1864) 'Lichenes in Aegypto a Ehrenbergi Collecti', *Act. Soc. Linn. Bordeaux,* 25 : 63.

Poelt J. (1962) 'Bestimmungsschlüssel der höheren Flechten von Europa', *Mitt. Bot. Staatssamml., Münch.,* 4 : 301–571.

— (1965) 'Zur Systematik der Flechtenfamilie *Physciaceae*', *Nova Hedwigia,* 9 : 21–32.

Reichert I. (1940) 'A New Species of *Diploschistes* from Oriental Steppes and its Phytogeographical Significance', *Palest. J. Bot.* (Rehovot Ser.), 3 : 162–182.

Runemark M. (1956) 'Studies in *Rhizocarpon,* I: Taxonomy of the Yellow Species in Europe', *Opera Botanica,* 2 : 5–152.

Verseghy K. (1962) 'Die Gattung *Ochrolechia*', Suppl., *Nova Hedwigia,* 1 : 1–146.

Weber W. A. (1962) 'Environmental Modification and the Taxonomy of the Crustose Lichens', *Svensk Bot. Tidskr.,* 56 : 293–333.

Wetmore C. M. (1960) 'The Lichen Genus *Nephroma* in North and Middle America', *Publ. Mus. Mich. St. Univ.* (Biol. Ser.), 1 : 373–452.

Zahlbruckner A. (1926) 'Lichenes (Flechten)', in: A. Engler & K. Prantl (eds.), *Die natürlichen Pflanzenfamilien* (2nd ed.), Engelmann, Leipzig.

Zohary M. (1966) *Flora Palaestina,* I: *Equisetaceae to Moringaceae,* The Israel Academy of Sciences and Humanities, Jerusalem.

INDEX

(Synonyms are given in italics)

Acarospora Mass. 44
  areolata Reichert et Galun 45
  bornmuelleri Stein. 45
  murorum Mass. 46
  reagens Zahlbr. f. radicans (Nyl.) Magn. 46
ACAROSPORACEAE 44
Amphiloma
  ehrenbergii Müll. Arg. *89*
  erythrinum var. *pulvinatum* Müll. Arg. *90*
Anaptychia Körb. 81
  ciliaris (L.) Körb. 81, 82
  intricata (Desfont.) Mass. *82*
Arthonia Ach. 103
  melanophthalma Duf. 103
ARTHONIACEAE 102
Arthopyrenia Mass. 104
  conoidea (E. Fr.) Zahlbr. 104
ARTHOPYRENIACEAE 103
Aspicilia Mass. 49
atranorin 37, 52, 53, 65, 66, 74, 76, 77, 78, 89
Bacidia De Not. 26
  albescens (Kremp.) Zw. 26
Baeomyces pocillum Ach. *36*
barbatic acid 33
Biatora ferruginea var. *festiva* E. Fr. *85*
Blastenia Mass. 83
  latzeli Serv. 83
  rejecta Th. Fr. var. bicolor (Müll. Arg.) Zahlbr. 83
  melanocarpa var. *bicolor* Müll. Arg. *83*
Buellia De Not. 72
  canescens (Dicks.) De Not. 74
  epipolia (Ach.) Mong. 73
  sorediosa Reichert et Galun 72
  subalbula (Nyl.) Müll. Arg. var. fusco-capitellata M. Lamb 73
  venusta (Körb.) Lett. 73
  zoharyi Galun 74
Callopisma
  conglomeratum Bagl. *91*
  interveniens Müll. Arg. *86*
Caloplaca Th. Fr. 84
  aegyptiaca (Müll. Arg.) Stein. var. circinans Stein. 88
  aurantia (Pers.) Hellb. var. aurantia Poelt 86, 89
  carphinea (Fr.) Jatta 89
  citrina (Hoffm.) Th. Fr. 85
  conglomerata (Bagl.) Jatta 91
  dalmatica (Mass.) Zahlbr. *83*
  ehrenbergii (Müll. Arg.) Zahlbr. 89
  erythrina (Müll. Arg.) Zahlbr. var. pulvinata (Müll. Arg.) Zahlbr. 90
  erythrocarpa (Pers.) Zw. 90
  festiva (E. Fr.) Zw. 85
  flageyana (Flag.) Zahlbr. 85, 86
  haematites (Chaub.) Zw. 86
  interveniens (Müll. Arg.) Zahlbr. 86
  lallavei (Clem.) Flag. *90*
  lamprocheila Flag. 87
  luteoalba (Turn.) Th. Fr. 87
  murorum (Hoffm.) Th. Fr. 91
  negevensis Reichert et Galun 91
  ochracea (Schaer.) Flag. 86, 92
  squamulosa (Wedd.) B. de Lesd. *91*
  tetrasticha (Nyl.) Oliv. *92*
  variabilis (Pers.) Müll. Arg. 87
  velana DR. 84
Candelariella Müll. Arg. 62
  medians (Nyl.) A.L. Sm. 62
  minuta Reichert et Galun 63
  vitellina (Ehrh.) Müll. Arg. 62
Carlosia Samp. 102
  lusitanica Samp. 102
Catillaria (Ach.) Th. Fr. 27
  chalybeia (Borr.) Mass. 27
    f. ilices (Mass.) Vain. 27
  piciloides Zahlbr. 27
  reichertiana Galun 27, 28
chloroatronorine 74

Cladonia Hill. 36
convoluta (Lam.) Cout. 36, 37
pocillum (Ach.) Rich. 36, 37, 39
rangiformis Hoffm. 37, 38
CLANDONIACEAE 36
Coccobotrys 97
COCCOCARPIACEAE 23
Coccomyxa 12, 24
Collema G. H. Web. 15
crispum (Huds.) G. H. Web. 16
cristatum (L.) G. H. Web. 16
nigrescens (Huds.) DC. 16
omphalarioides Anzi *18*
polycarpon Hoffm. 17
var. *corcyrense* *17*
pulposum
var. *vulgare* Schaer. *17*
α vulgare papulosum Schaer. *18*
tenax (Sw.) Ach. var. vulgare (Schaer.) Degel. 17
f. papulosum (Schaer.) Degel. 18
f. vulgare Degel. 18
tunaeforme (Ach.) Ach. 18
COLLEMATACEAE 15
CYPHELIACEAE 102
Cystococcus 33, 57
DERMATOCARPACEAE 98
Dermatocarpon (Eschw.) Th. Fr. 98
bucekii Nadv. et Serv. 99
convexum Reichert et Galun 99
desertorum Tom. 100
hepaticum (Ach.) Th. Fr. 100
lachneum 100
miniatum (L.) Mann 100
monstrosum (Schaer.) Wain. 101
rufescens 100
subcrustosum (Nyl.) Zahlbr. 101
Dichotrix orsiniana 25
DIPLOSCHISTACEAE 37
Diploschistes Norm. 38
actinostomus Zahlbr. 39
bryophilus (Erht.) Zahlbr. 39
calcareus Stein. 40
ocellatus Norm. 40
scruposus Norm. 41
steppicus Reichert 41, 42
tenuis Reichert et Galun *40*
diploschistic acid 39, 42
Diplotomma venustum *73*
Dirina E. Fr. 105
ceratoniae (Ach.) E. Fr. 105
DIRINACEAE 105
Endocarpon
cinerascens var. *subcrustosum* Nyl. *101*
hepaticum Ach. *100*
miniatum var. *monstrosum* Schaer. *101*
erythrin 32, 106
Eulecanora Th. Fr. 51
Eulecidea Stzbgr. 28
Evernia Ach. 66
prunastri (L.) Ach. 66
evernic acid 66, 69
Fulgensia Mass. et De Not. 92
fulgens (Sw.) Elenk. 92
fulgida (Nyl.) Szat. 93
subbracteata (Nyl.) Poelt 93
fumarprotocetraric acid 36, 37
gangaleoidin 53
Gasparrina (Tornab.) Th. Fr. 88
Gloeocapsa 12, 20, 21
glomelliferic acid 64
Gonohymenia Stein. 20
mesopotamica Stein. 20
Gyalolechia cinnabarina Flag. *85*
gyrophoric acid 89
Heppia Naeg. 22
psammophila Nyl. 22
HEPPIACEAE 22
Hyalococcus 98
Hypomorpha 23
Lecania Mass. 48
erysibe Mudd. 48
koerberiana Lahm. 48
nyderlandiana Mass. 49
subcaesia (Nyl.) Szat. 49
Lecanora Ach. 49
atra Ach. 28, 52
bolcana (Poll.) Poelt 55
ceratoniae Ach. *105*
contorta (Hoffm.) Stein. 51
crenulata (Dicks.) Hook. 52
desertorum Kphbr. 50
elegans var. *ectaniza* Nyl. *95*
farinosa Nyl. 51
gangaleoides Nyl. 53

haematites Chaub. *86*
hageni Ach. 53
hoffmannii Müll. Arg. 51
microspora Zahlbr. 50
muralis (Schreb.) Rabh. 55
olea Reichert et Galun 53
pruinosa Chaub. 56
radiosa (Hoffm.) Schaer. var. subcircinata (Nyl.) Zahlbr. 56
schleicheri dealbata f. *radicans* Nyl. *46*
scrupulosa Ach. 54
subcaesia Nyl. 49
subcircinata Nyl. *56*
subplanata Nyl. 54
subrugosa Nyl. 54
LECANORACEAE 47
lecanoric acid 39, 40, 41, 57, 64, 65
Lecidea Ach. 28
albilabra Duf. 30
algeriensis (Flag.) Zahlbr. 29
aromaticus Turn. *34*
bolcana Poll. *55*
decipiens (Ehrh.) Ach. 31
epipolia Ach. *73*
euphorea (Floerk.) Nyl. 29
ilices Mass. *27*
maculosa Flag. *29*
ocellulata Th. Fr. 29
ochracea (Schaer.) Flag. 92
olivacea (Hoff.) Mass. 30
opaca Duf. 31
parasema var. *elaeochromoides* Nyl. *30*
sabuletorum var. *euphorea* Floerk. *29*
subincongrua Nyl. var. elaeochromoides (Nyl.) Poelt 30
LECIDEACEAE 26
Lichen
aipolius Ehrh. *75*
bryophilus Erht. *39*
calvus Dicks. *32*
canescens Dicks. *74*
ciliaris L. *81*
coeruleonigricans Lightf. *34*
convolutus Lam. *36*
crassus Huds. *61*
crenulatus Dicks. *52*
crispus Huds. *16*
cristatus L. *16*
decipiens Ehrh. *31*
farinaceus L. *68*
fastigiatus Pers. *67*
fulgens Sw. *92*
gypsaceus Sm. *61*
immersum Web. *32*
lentigerus Web. *60*
leptalea Ach. *78*
luteoalbus Turn. *87*
miniatus L. *100*
murorum Hoffm. *91*
muralis Schreb. *55*
nigrescens Huds. *16*
parellus L. *57*
parietinus L. *95*
pollinarius Ach. *69*
pruinosus Sm. *47*
prunastri L. *66*
pulverulentus Schreb. *80*
rupicola Hoffm. *51*
stellaris L. *76*
tenellus Scop. *77*
tiliaceus Hoffm. *65*
variabilis Pers. *87*
vitellinus Ehrh. *62*
Myrmecia 12, 97, 98
biatorella 47
Nephroma Ach. 24
laevigatum Ach. 24
nephromin 24
norstictic acid 31, 40, 44, 47, 57, 68, 69
Nostoc 12, 15, 18, 22, 23, 24
Ochrolechia Mass. 57
parella (L.) Mass. 57
olivetoric acid 54
Opegrapha Humb. 104
atra Pers. 104
grumulosa Duf. 105
OPEGRAPHACEAE 104
PANNARIACEAE 22
Pannaria Del. 23
mediterranea C. Tav. 23
pannarin 58
parietin 85, 86, 87, 89, 91, 92, 93, 94, 96
Parmelia Ach. 63
aureola Ach. *95*
carphinea Fr. *89*

farrea Ach. *81*
glabra Nyl. 63
glomellifera Nyl. 64
maciformis Del. *67*
perrugata Nyl. 64
pulverulenta f. *lilacina* Arn. *79*
tiliacea (Hoffm.) Ach. 65
tinctina Mah. et Gill. 65
tunaeformis Ach. *18*
venusta Ach. *80*
PARMELIACEAE 63
Patellaria erythrocarpa Pers. *90*
Peccania (Mass.) Forss. 20, 21
PELTIGERACEAE 24, 25
Pertusaria DC. 42
carmeli Reichert et Galun 42, 43
ilicicola Harm. 43
leucostoma (Bernh.) Mass. var. areolascens Erichs. 43
multipuncta (Turn.) Nyl. var. leptosporoides Erichs. 44
PERTUSARIACEAE 42
Physcia (Schreb.) DC. 75
aipolia (Ehrh.) Hampe 75
ascendens Bitt. 77, 78
biziana (Mass.) Zahlbr. 76
leptalea (Ach.) DC. 78
stellaris (L.) Nyl. 76
subpulverulenta Szat. *80*
tenella(Scop.) Bitt. 77
PHYSCIACEAE 69
Physconia Poelt 79
farrea (Ach.) Poelt 81
grisea (Lam.) Poelt ssp. lilacina (Arn.) Poelt 79
pulverulenta (Schreb.) Poelt 80
subpulverulenta (Szat.) Poelt 80
venusta (Ach.)Poelt 80
Physma Mass. 18
omphalarioides (Anzi) Arn. 18, 19
Placodium (Pers.) Poelt 55
Placodium Pers.
fulgidum Nyl. *93*
medians Nyl. *62*
PLACYNTHIACEAE 25
Placynthium (Ach.) S. F. Gray 25
nigrum (Huds.) S. F. Gray 25
Pleurococcus 12, 62, 97
Protoblastenia Stein. 32
calva (Dicks.) Zahlbr. 32
immersa (Web.) Stein. 32
protocetraric acid 69
Psora 30
psoromic acid 33, 60, 61
Psorotichia Mass. 21
numidella Forss. 21
PYRENOPSIDACEAE 19
Ramalina Ach. 66
duriaei (De Not.) Jatta 68
farinacea (L.) Ach. 68, 69
farinacea var. *reagens* B. de Lesd. *69*
fastigiata (Pers.) Ach. 67
lacunosa Rupr. *94*
maciformis (Del.) Bory 67, 68
pollinaria (Ach.) Ach. 69
pollinaria var. *duriaei* De Not. *68*
reagens (B. de Lesd.) Culb. 69
subfarinacea 69
Rhizocarpon Ram. 33
tinei (Tornab.) Run. ssp. tinei Run. 33
Rinodina (Ach.) S. Gray 70
bischoffii (Hepp) Mass. var. aegyptiaca Müll. Arg. 71
carmeli Reichert et Galun 71
magnussoniana Reichert et Galun 72
mediterranea Flag. 71
salazinic acid 40, 44, 57, 65, 69
Sarcogyne Flotow 47
pruinosa (Sm.) Körb. 47
Scoliciosporum molle f. *albescens* *26*
Scytonema 12, 22
Solenopsora Mass. 58
candicans Stein. 58
cesatii Zahlbr. var. grisea Bagl. 58
montagnei Choisy et Werner var. calcarea Schaer. 59
Spilonema Born. 23
revertens Nyl. 23
Squamaria biziana Mass. *76*
Squamarina Poelt 59
crassa (Huds.) Poelt 61, 85
var. crassa f. crassa Poelt 61
f. pseudocrassa (Matt.) Poelt 61
f. iberica (Matt.) Poelt 62
gypsacea (Sm.) Poelt 61

lentigera (Web.) Poelt 60
var. *pseudocrassa* Matt. *61*
stella-petraea Poelt 60
stictaurin 62, 63
stictic acid 74
Stigonema 12, 23
TELOSCHISTACEAE 83
Teloschistes Norm. 94
brevior f. *halophilus* (Elenk.) Oxn. *94*
lacunosus (Rupr.) Sav. 94
Thalloidima verrucosum Mass. *35*
Thyrea Mass. 21, 22
Toninia Th. Fr. 33
albomarginata B. de Lesd. 33
aromatica (Turn.) Mass. 34
coeruleonigricans (Lightf.) Th. Fr. 34
verrucosa (Mass.) Flag. 35
Tornabenia Mass. 82
intricataTrevis. 82
Trebouxia 12, 28, 36, 49, 63, 66, 69, 70, 72, 75, 79, 83, 84, 94
decolorans 95
Trentepohlia 12, 102, 103, 104, 105
Triophthalmidium Zahlbr. 92
USNEACEAE 66
usnic acid 28, 36, 53, 54, 55, 56, 60, 65, 66, 67, 69, 89
variolaric acid 57
Verrucaria (Wigg.) Th. Fr. 97
citrina Hoffm. *85*
conoidea E. Fr. *104*
fuscella Ach. 97
marmorea Ach. 97
olivacea Hoffm. *30*
viridula Ach. 98
VERRUCARIACEAE 96
Xanthoria Th. Fr. 94
aureola (Ach.) Erichs. 95
var. aureola Poelt 95
var. ectaniza (Nyl.) Poelt 95
var. isidioidea Beltr. 96
parietina (L.) Th. Fr. 86, 95
steineri Lamb 86, 96
zeorin 58

1. *Collema nigrescens* (× 2.5)

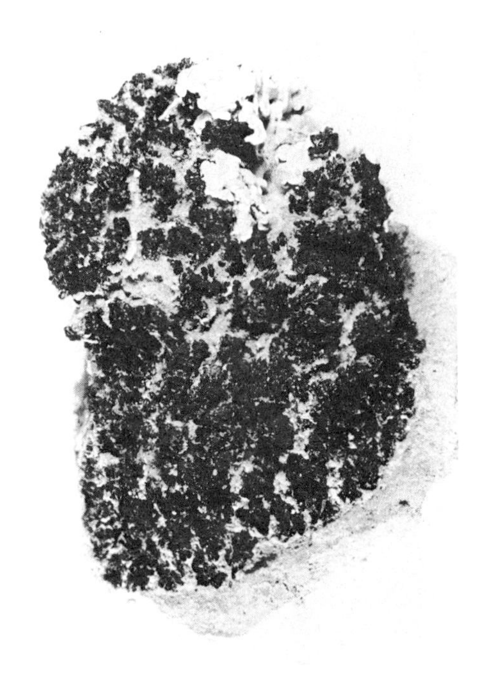

2. *Collema tenax* v. *vulgare* f. *vulgare* (× 2)

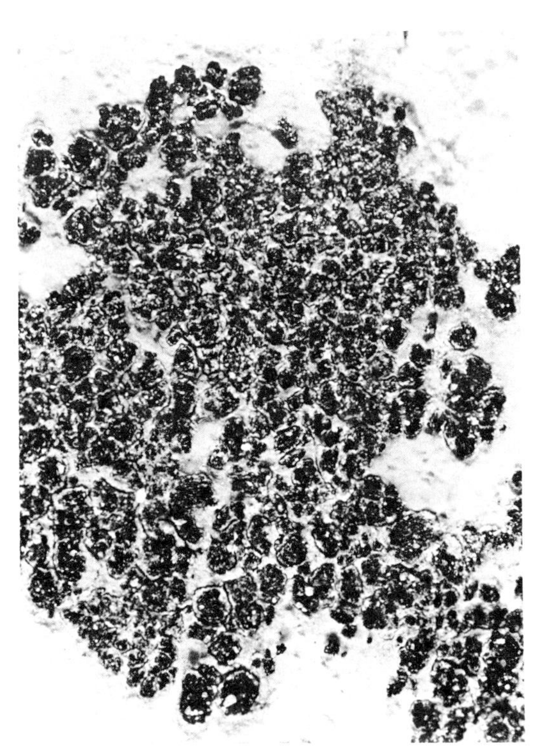

3. *Psorotichia numidella* (× 10)

4. *Physma omphalarioides* (× 1.75)

PLATE II

1. *Gonohymenia mesopotamica* (× 4.5)
(from Galun, 1967b)

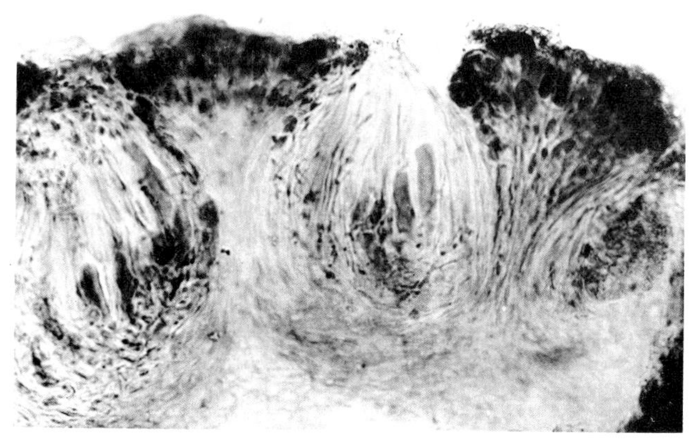

2. *Gonohymenia mesopotamica*
Two hymenia separated by a sterile zone (× 250)
(from Galun, 1967b)

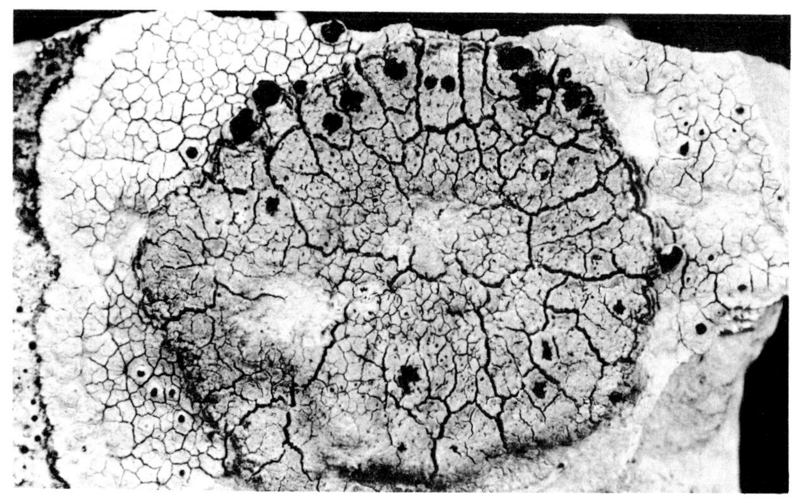

3. *Catillaria reichertiana* (× 1.2)
(from Galun, 1967a)

1. *Lecidea euphorea* (× 6)

2. *Lecidea olivacea* (× 3.5)

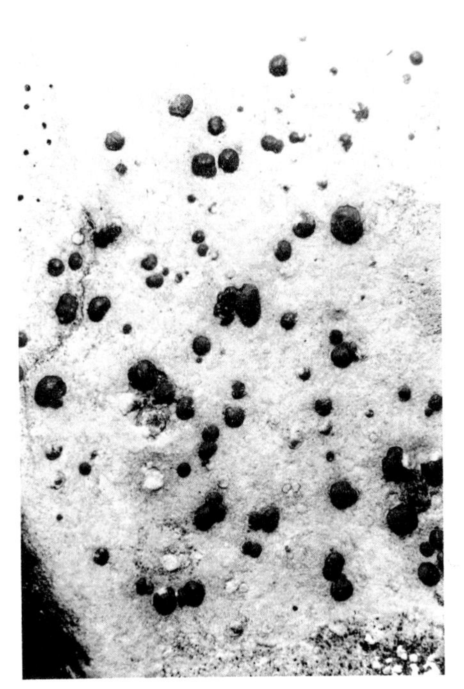

3. *Protoblastenia calva* (× 3)

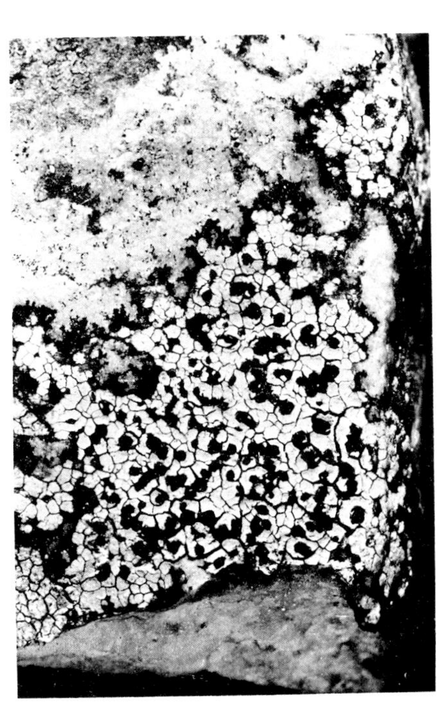

4. *Rhizocarpon tinei* ssp. *tinei* (× 3)

PLATE IV

1. *Lecidea albilabra* (× 8)

2. *Lecidea decipiens* from the Negev (× 7)

3. *Lecidea decipiens* from Upper Galilee (× 0.5)

PLATE V

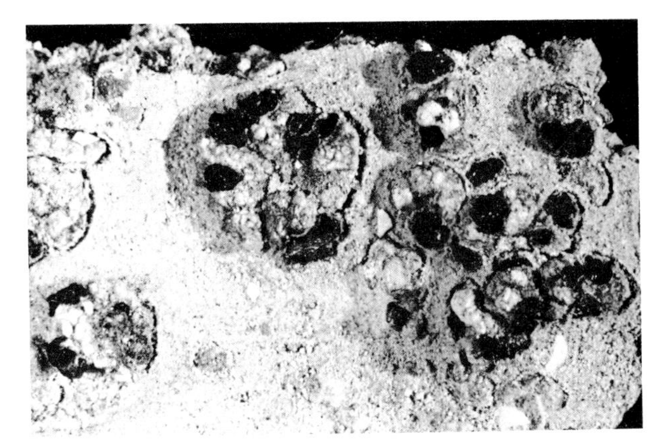

1. *Toninia aromatica* (× 5)

2. *Toninia albomarginata* (× 5)

3. *Toninia coeruleonigricans* (× 1.5)
(from Galun, 1966)

1. *Cladonia convulata* (× 2)

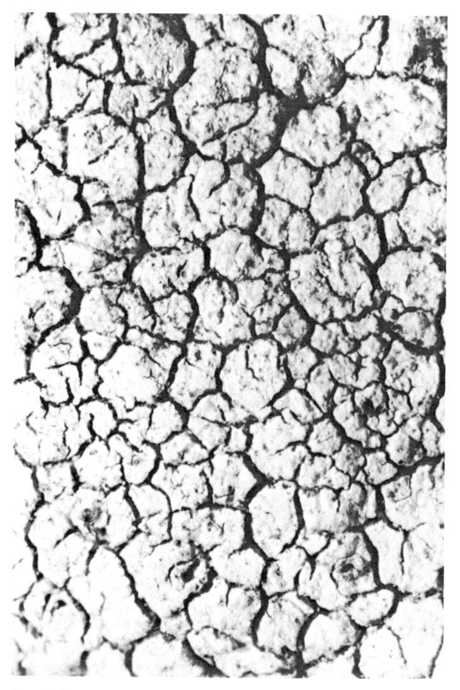

2. *Diploschistes actinostomus* (× 9)

3. *Diploschistes bryophilus* (× 9.5)

1. *Diploschistes steppicus* (× 5)

2. *Diploschistes ocellatus* (× 3.2)

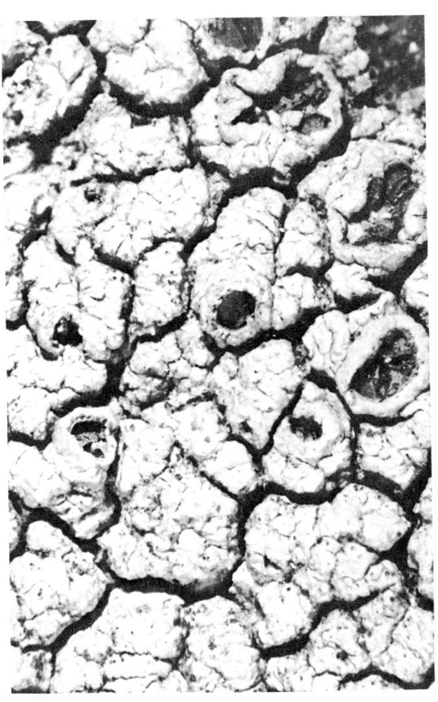

3. *Diploschistes scruposus* (× 6)

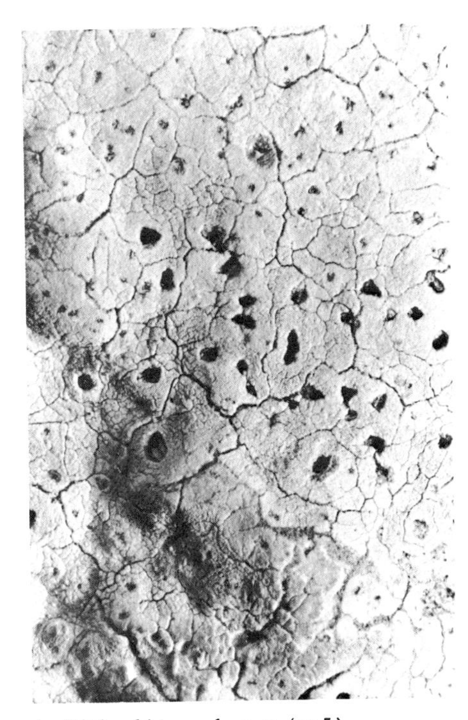

4. *Diploschistes calcareus* (× 5)

PLATE VIII

1. *Pertusaria ilicicola* (× 3.5)

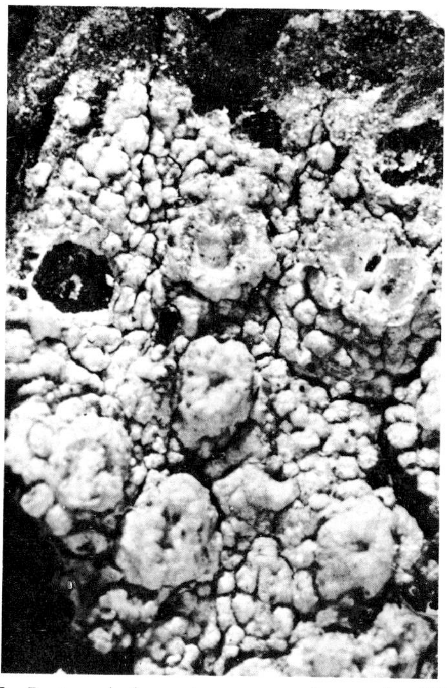

2. *Pertusaria leucostoma* v. *areolascens* (× 10)

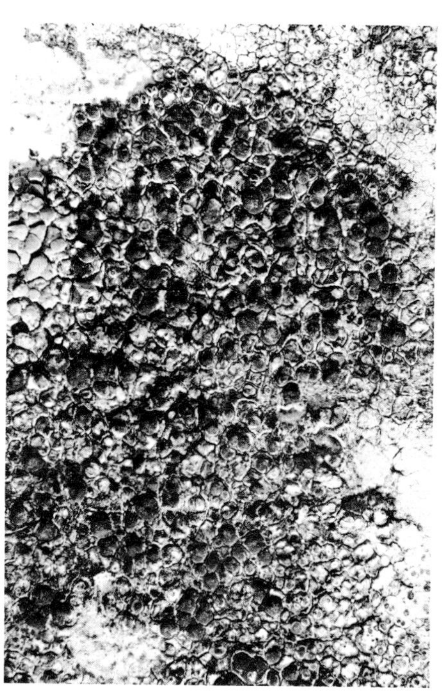

3. *Lecania erysibe* (× 4)

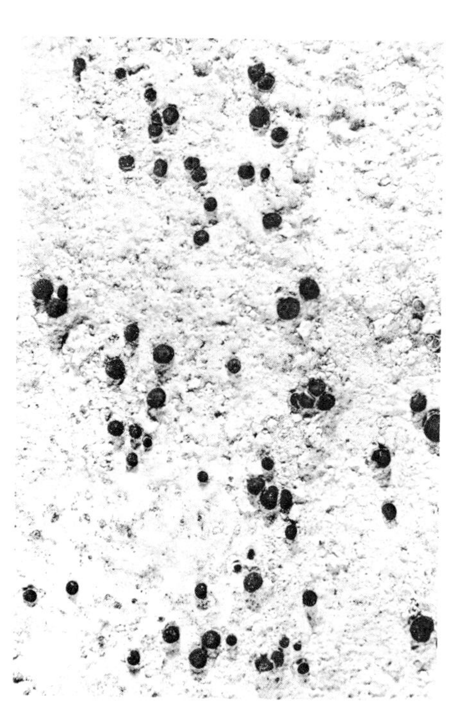

4. *Lecania subcaesia* (× 4)

1. *Acarospora murorum* (× 7)

2. *Acarospora areolata* (× 8)

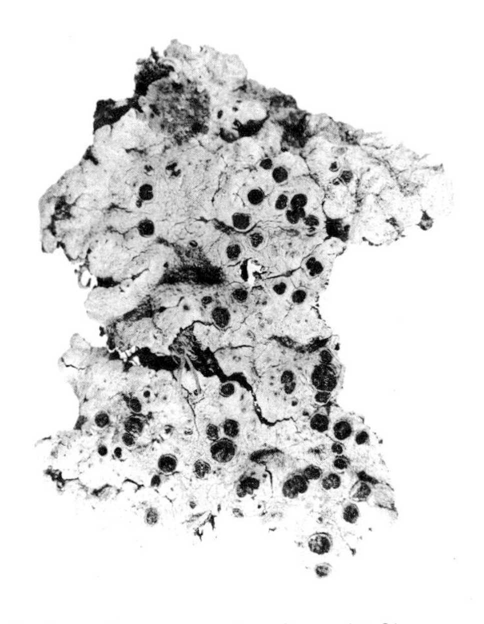

3. *Acarospora reagens* f. *radicans* (× 2)

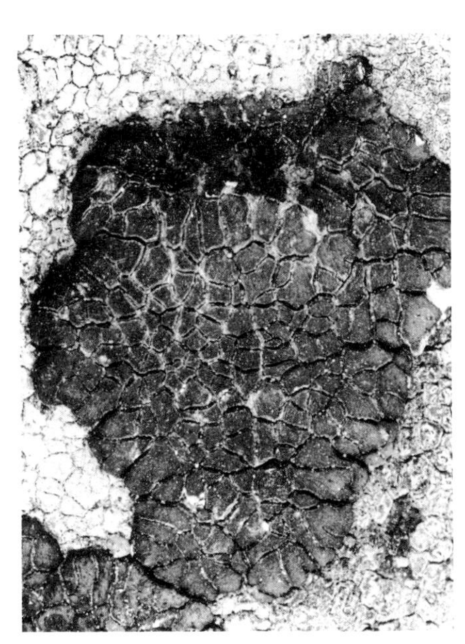

4. *Acarospora bornmülleri* (× 5.5)

PLATE X

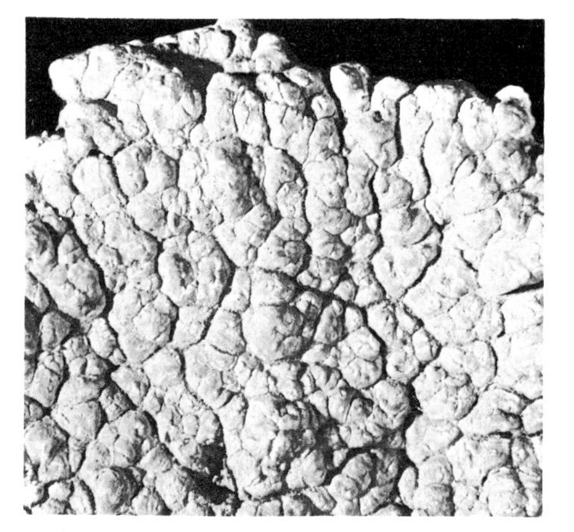

1. *Lecanora desertorum* (× 3)

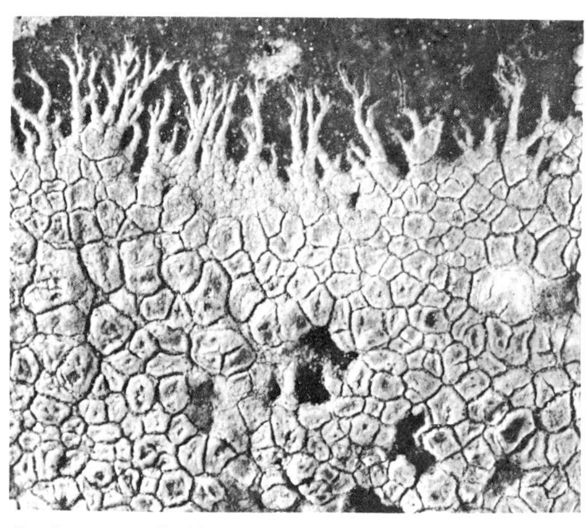

2. *Lecanora hoffmannii* (× 8)

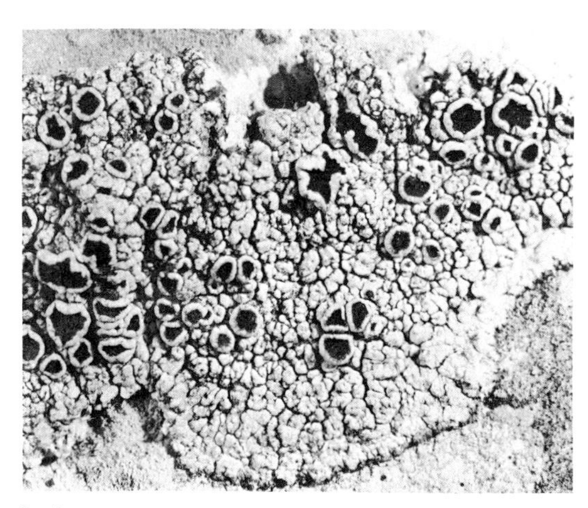

3. *Lecanora atra* (× 1.8)

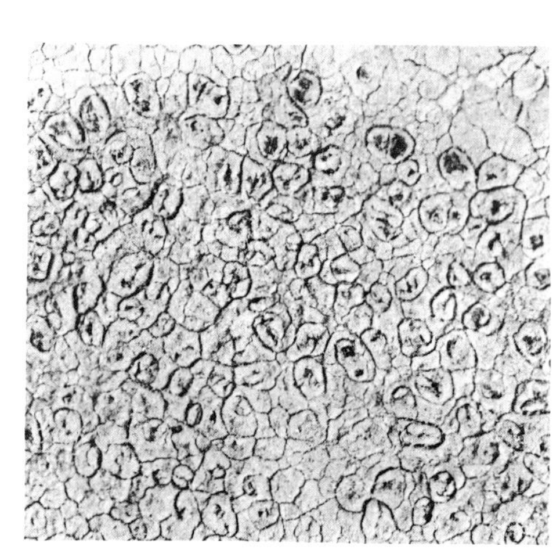

4. *Lecanora farinosa* (× 5.5)

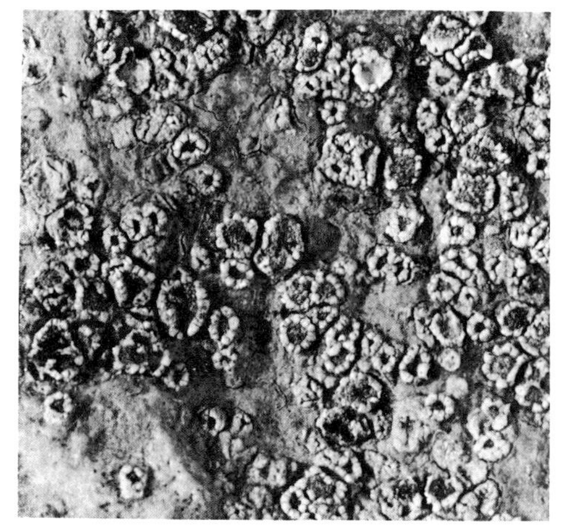

5. *Lecanora crenulata* (× 6.5)

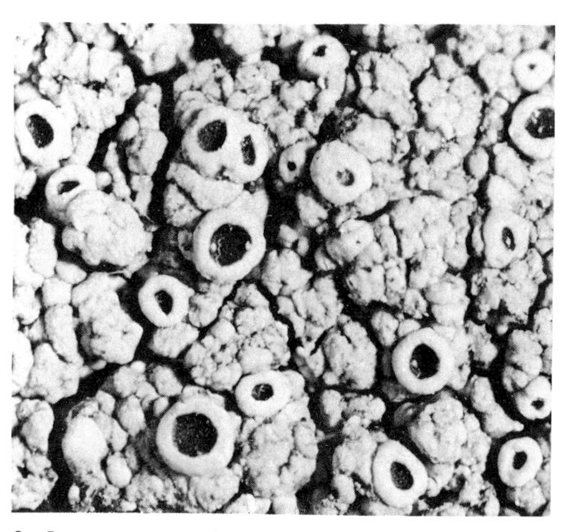

6. *Lecanora gangaleoides* (× 10.5)

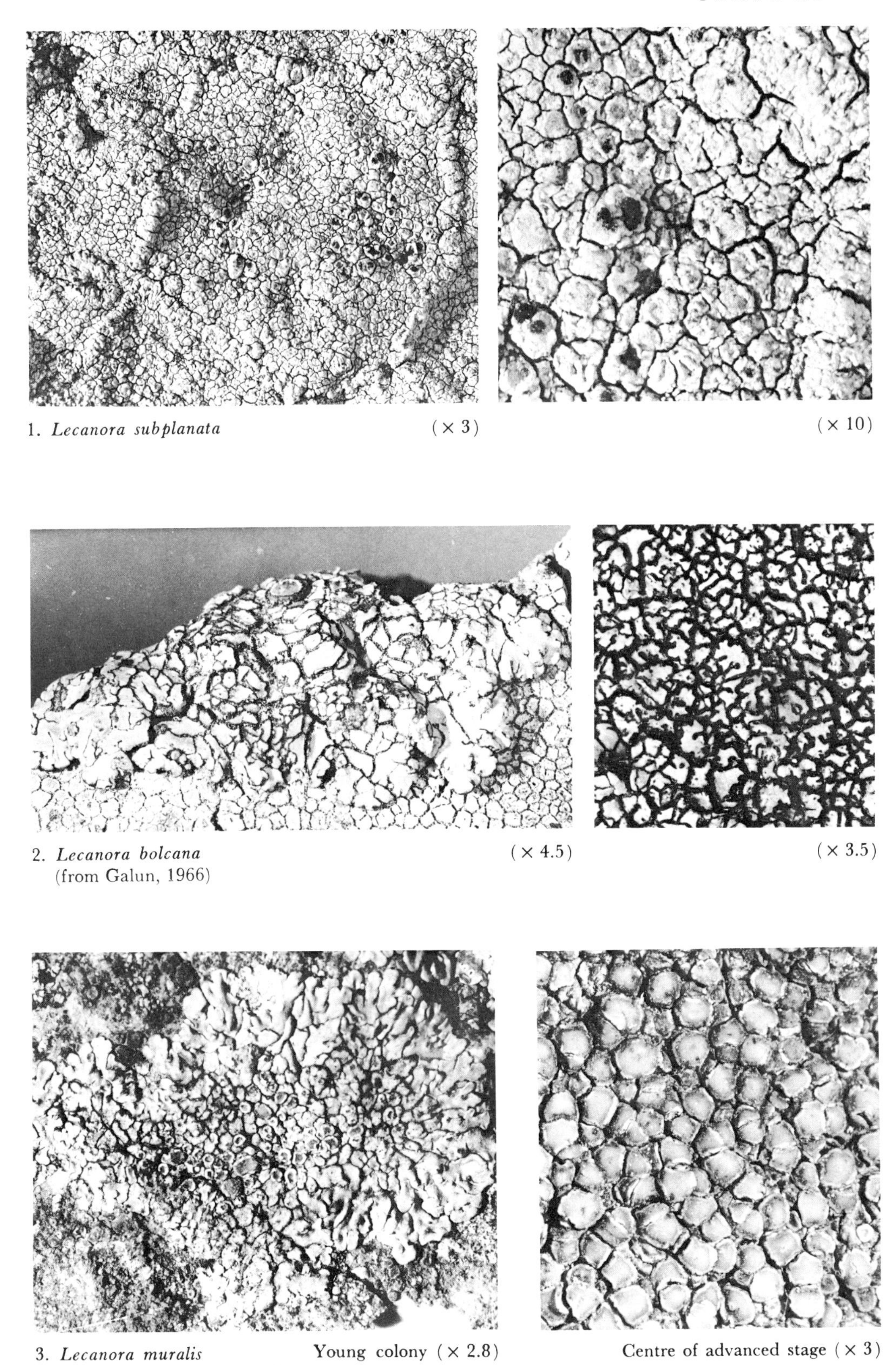

1. *Lecanora subplanata* (× 3) (× 10)

2. *Lecanora bolcana* (× 4.5) (× 3.5)
(from Galun, 1966)

3. *Lecanora muralis* Young colony (× 2.8) Centre of advanced stage (× 3)
(from Galun & Lavee, 1966)

PLATE XII

1. *Lecanora olea* (× 4.5)

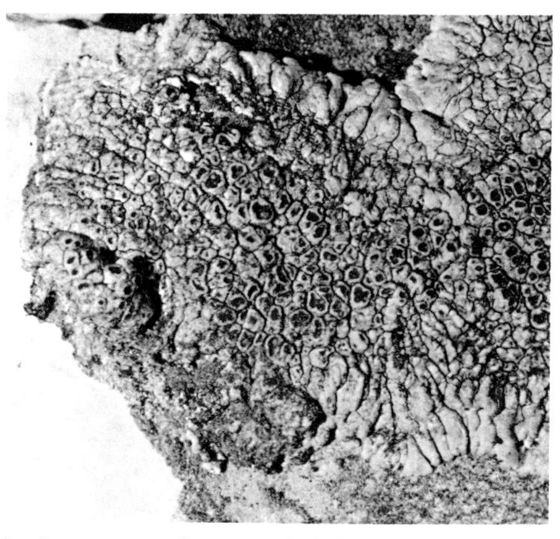

2. *Lecanora radiosa* v. *subcircinata* (× 2)

3. *Lecanora pruinosa* (× 4.5)

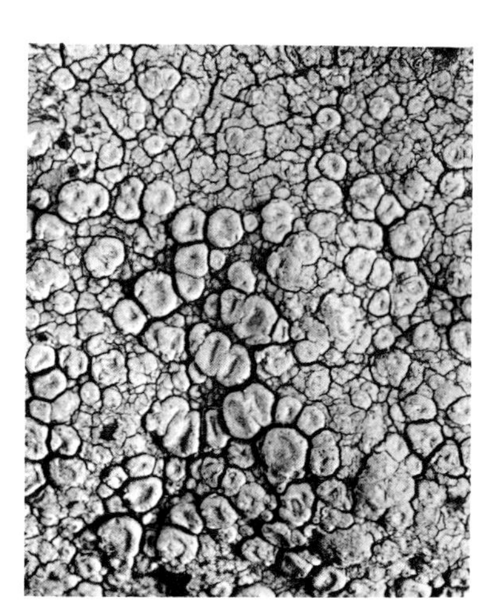

4. *Ochrolechia parella* (× 3)

5. *Candelariella medians* (× 1.5)

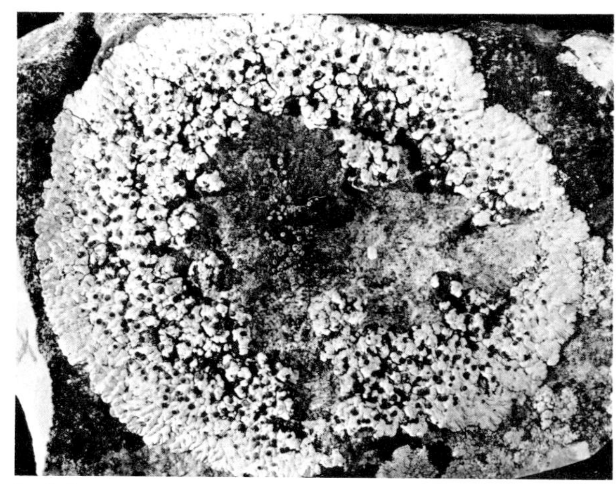

6. *Solenopsora candicans* (× 1)

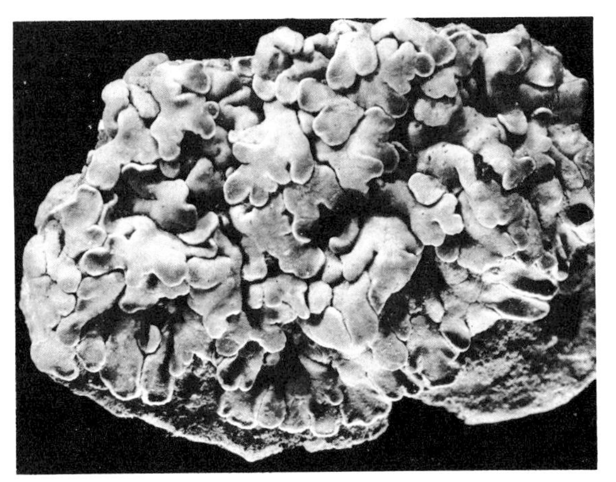

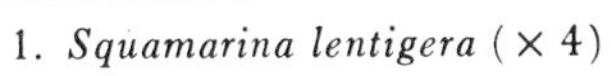

1. *Squamarina lentigera* (× 4)

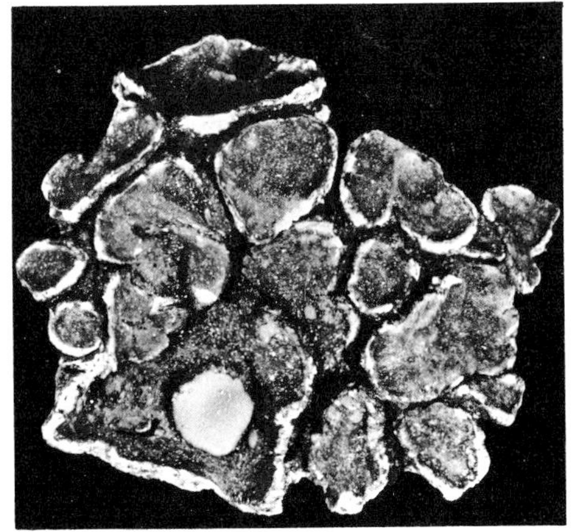

2. *Squamarina gypsacea* (× 4)

3. *Squamarina stella-petraea* (× 0.8)

Lobate zone

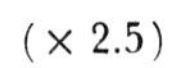

(× 2.5)

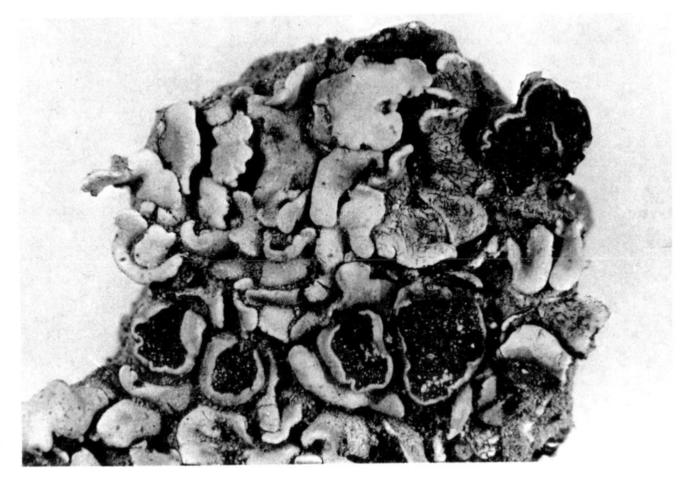

4. *Squamarina crassa* v. *crassa* f. *crassa* (× 4.7)

PLATE XIV

1. *Parmelia glabra* (× 2.5)

2. *Parmelia tiliacea* (× 2)
(from Galun, 1966)

3. *Parmelia perrugata* (× 15)

4. *Parmelia glomellifera* (× 2.8)

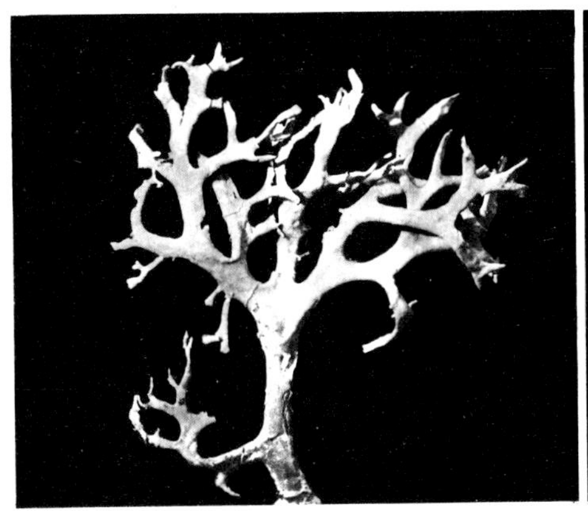

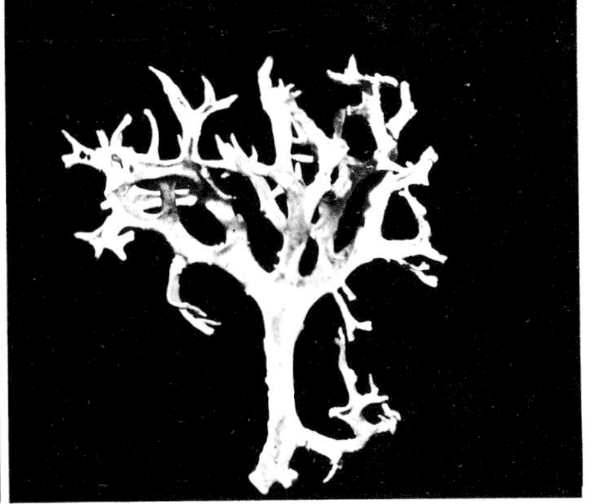

1. *Evernia prunastri* (× 2) Upper side Lower side

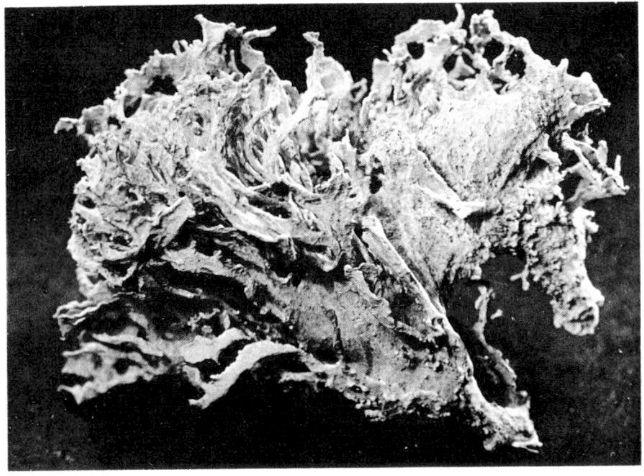

2. *Ramalina macifomis* (× 1.5)

3. *Ramalina fastigiata* (× 2.5)

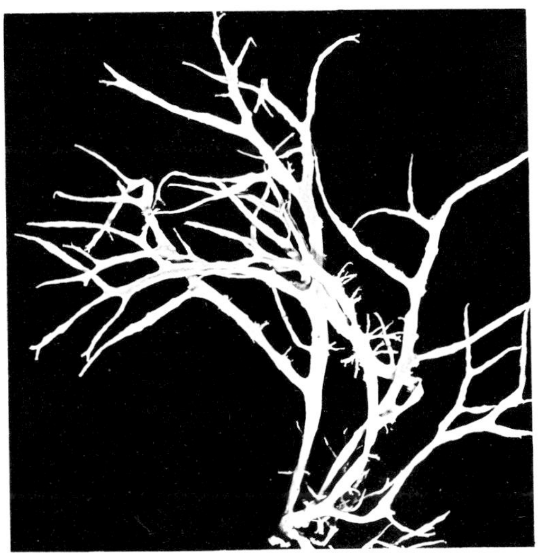

4. *Ramalina farinacea* (× 2)

5. *Ramalina pollinaria* (× 10)

PLATE XVI

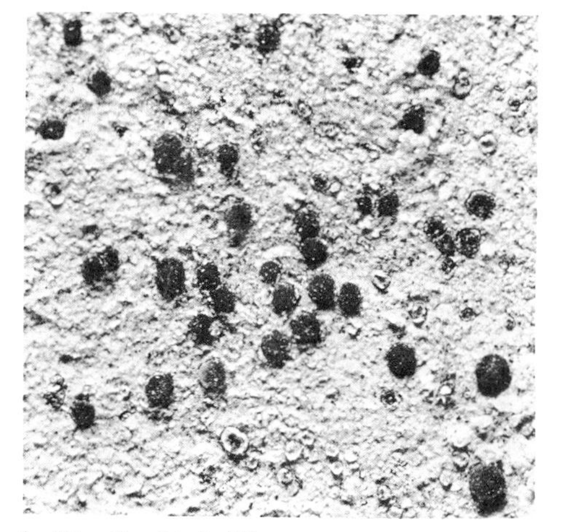

1. *Rinodina bischoffii* v. *aegyptiaca* (× 6)

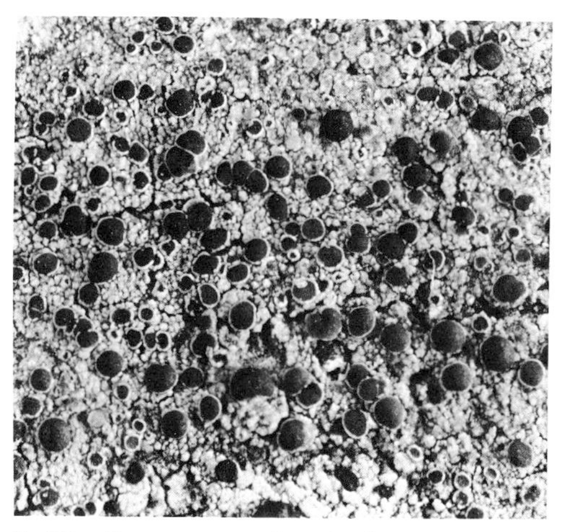

2. *Rinodina magnussoniana* (× 7)

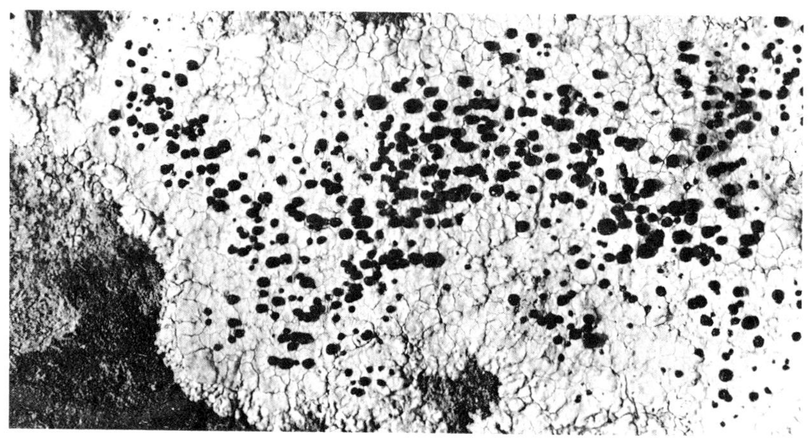

3. *Buellia subalbula* v. *fuscocapitellata* (× 4.5)

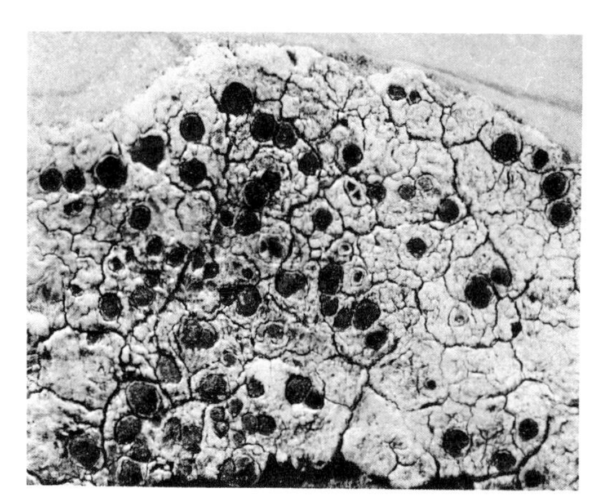

4. *Buellia venusta* (× 5.5)

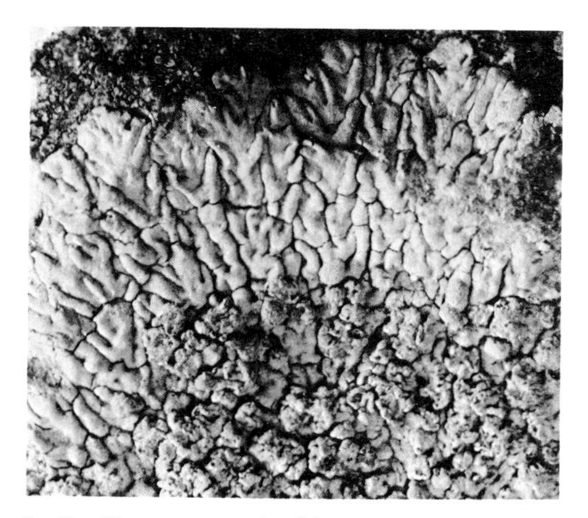

5. *Buellia canescens* (× 5)

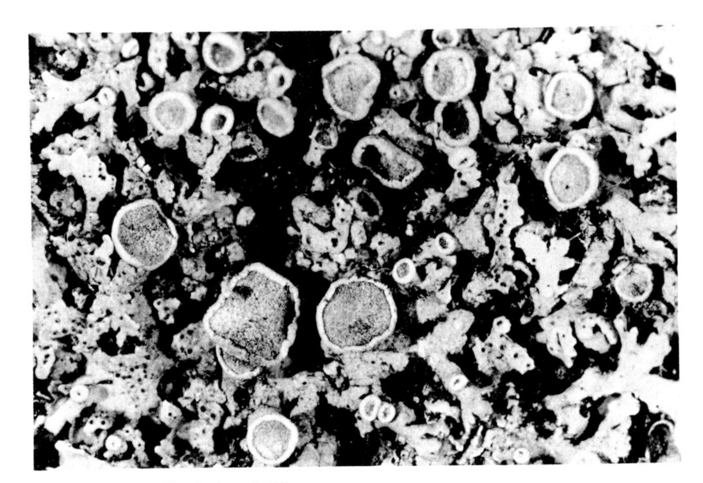

1. *Physcia stellaris* (× 6.5)

2. *Physcia ascendens* (× 8)

3. *Physcia biziana* (× 3.6)

PLATE XVIII

1. *Physconia farrea* (× 4)

2. *Physconia venusta* (× 5.5)

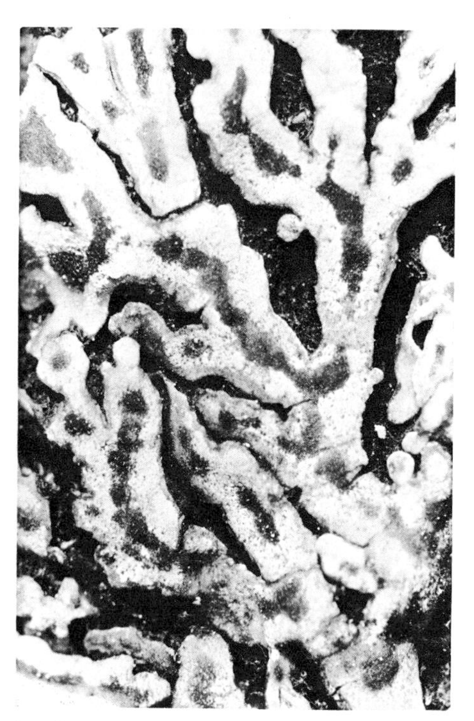

3. *Physconia pulverulenta* (× 11.5)

4. *Tornabenia intricata* (× 3)

PLATE XIX

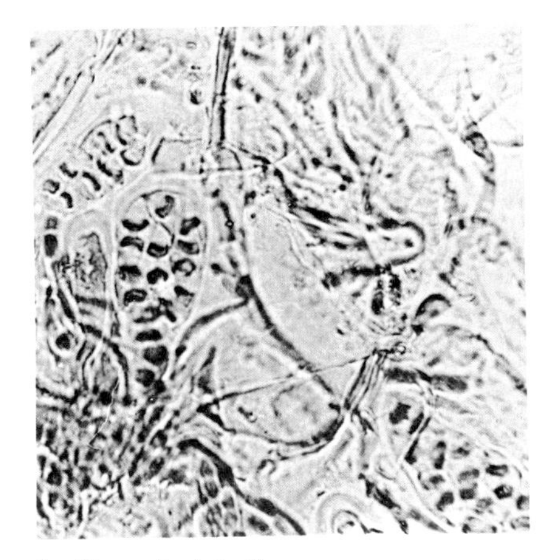

1. *Blastenia latzeli*
mischoblastiomorph spores (× 1000)

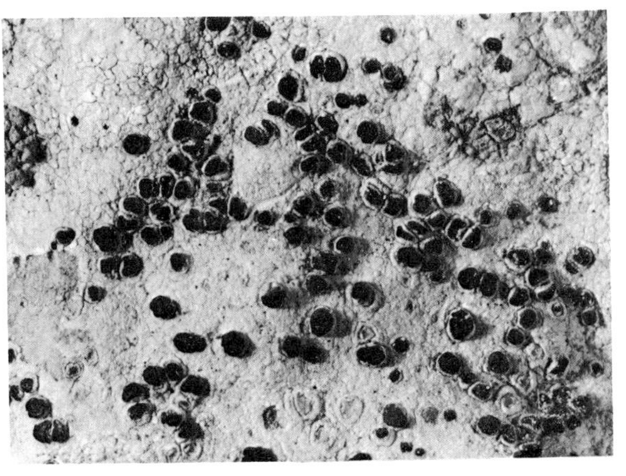

2. *Blastenia rejecta* v. *bicolor* (× 5)

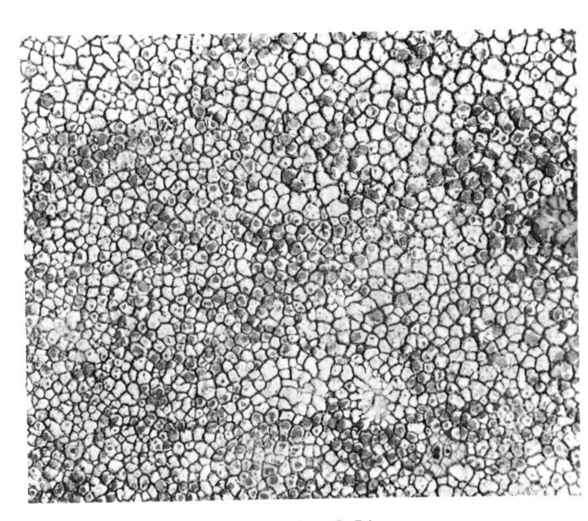

3. *Caloplaca carphinea* (× 2.5)

4. *Caloplaca luteoalba* (× 7)

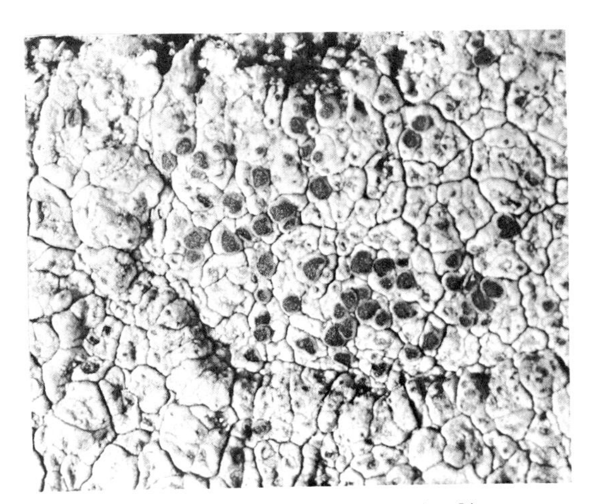

5. *Caloplaca aegyptiaca* v. *circinans* (× 2)

6. *Caloplaca festiva* (× 5.5)

PLATE XX

1. *Caloplaca aurantia* v. *aurantia* (× 3)

2. *Caloplaca ochracea* (× 14)

3. *Caloplaca ehrenbergii* (× 5.5)

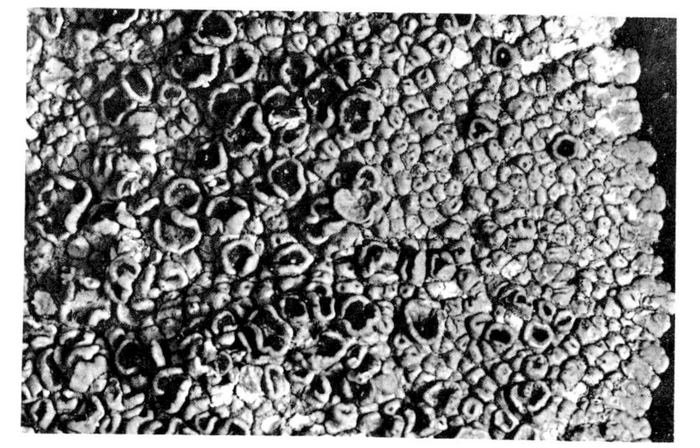

3. *Caloplaca ehrenbergii* (× 3)

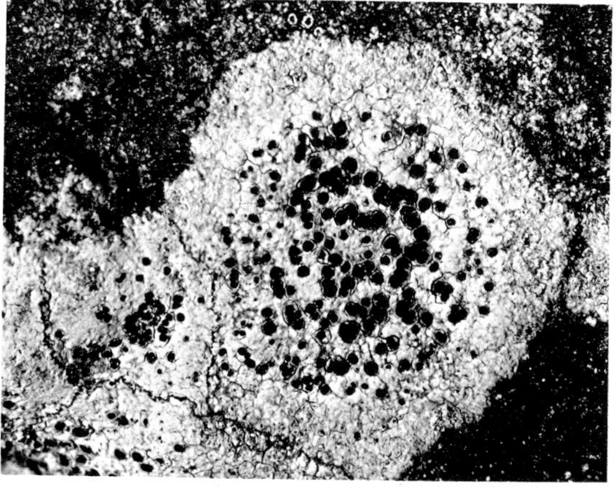

4. *Caloplaca erythrocarpa* (× 5)

1. *Xanthoria aureola* v. *ectaniza* (× 5)

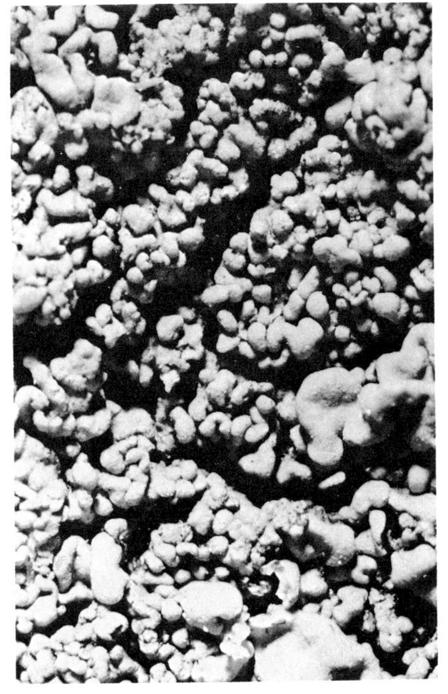

2. *Xanthoria aureola* v. *aurcola* (× 10)

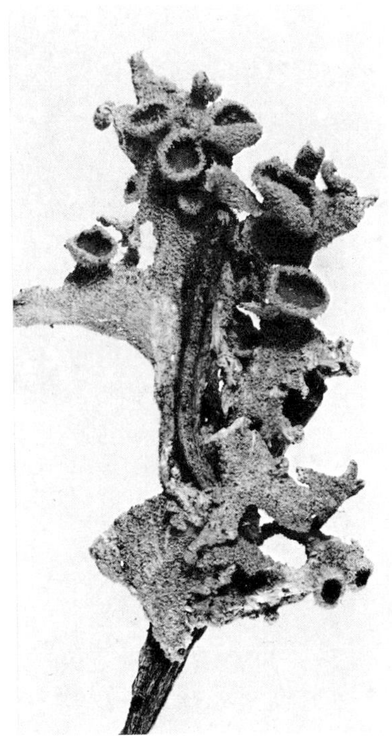

3. *Teloschistes lacunosus* (× 2.5)

4. *Xanthoria aureola* v. *isidioidea* (× 5.5)

PLATE XXII

1. *Fulgensia fulgida* (× 2)

2. *Fulgensia subbracteata* (× 6)

3. *Verrucaria fuscella* (× 4)

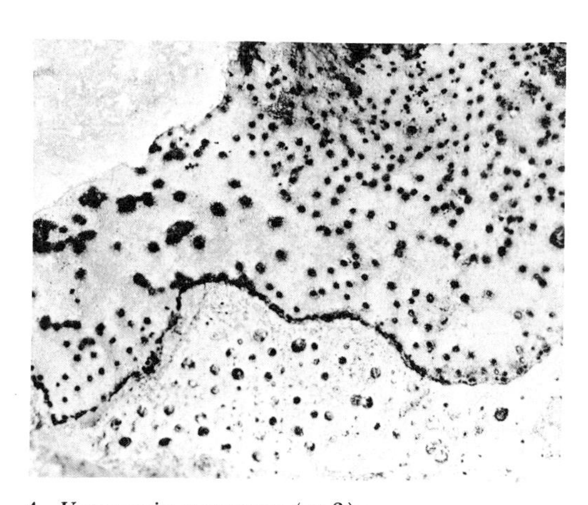

4. *Verrucaria marmorea* (× 3)
(from Galun, 1966)

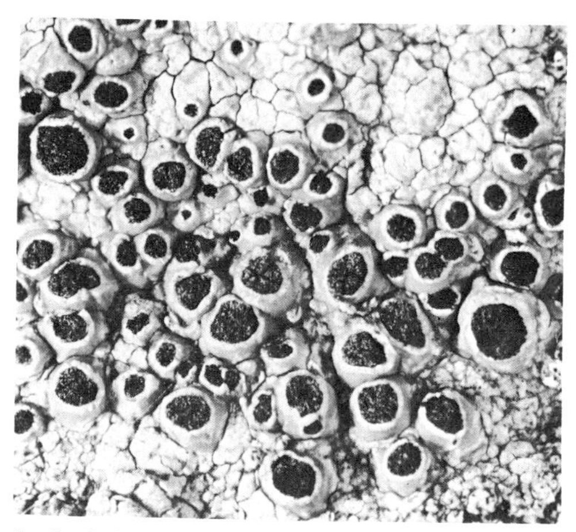

5. *Carlosia lusitanica* (× 6)

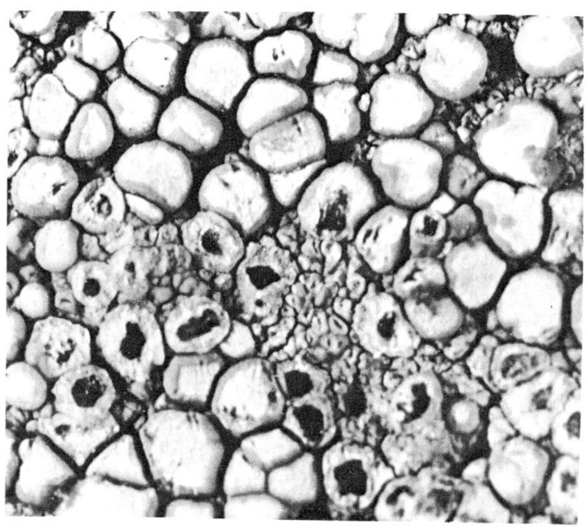

6. *Dirina ceratoniae* (× 6.5)

PLATE XXIII

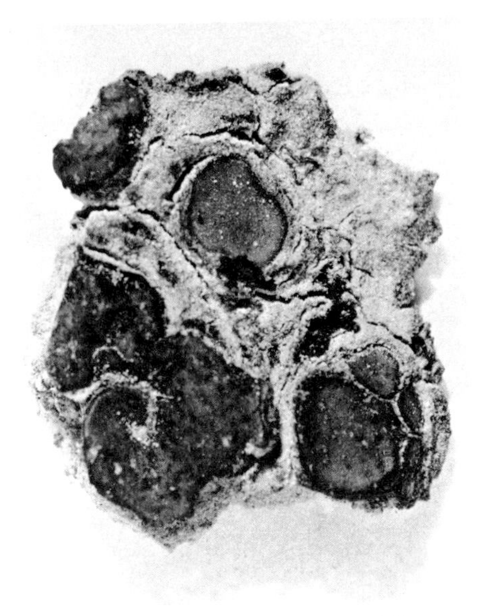

1. *Dermatocarpon hepaticum* (× 4.5)

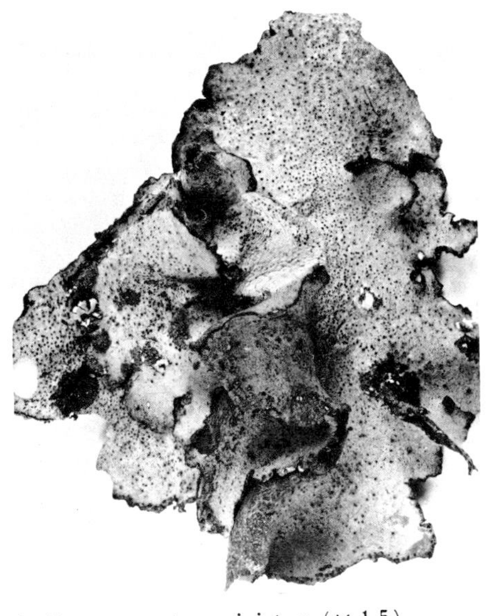

2. *Dermatocarpon miniatum* (× 1.5)

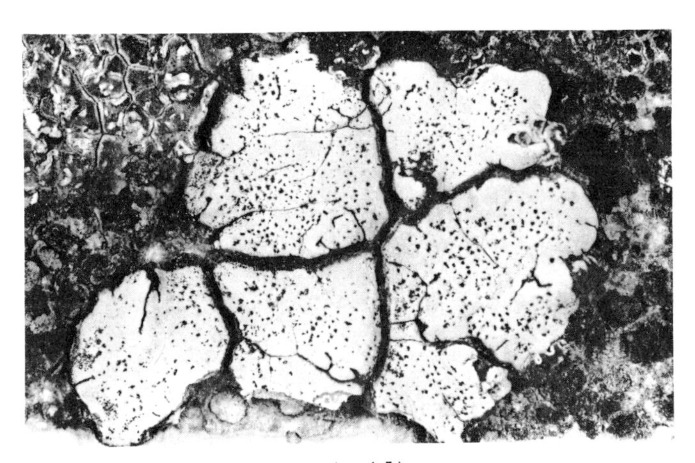

3. *Dermatocarpon monstrosum* (× 4.5)

4. *Dermatocarpon bucekii* (× 2)

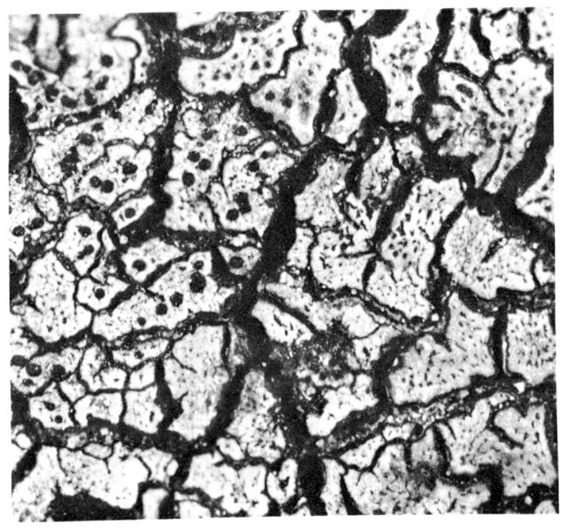

5. *Dermatocarpon subcrustosum* (× 6)

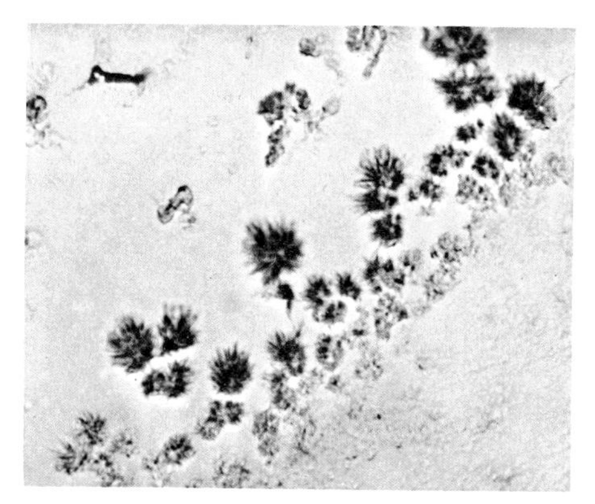

1. Protocetraric acid, in GAW (× 500)
(from Galun & Lavee, 1966)

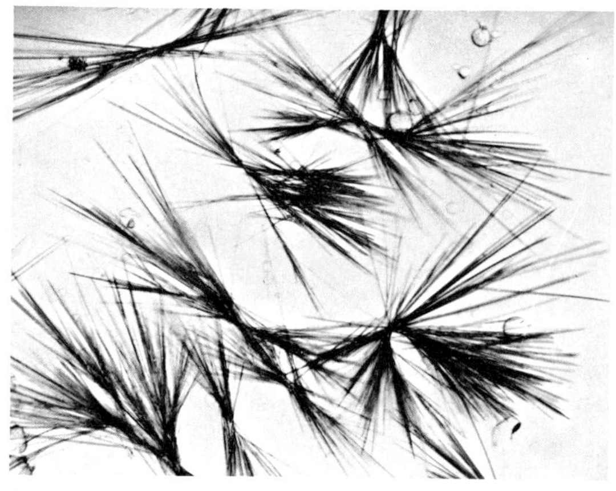

2. Unknown substance, in GAAn (× 125)

3. Erythrin, in GAW (× 450)

4. Barbatic and psoromic acid, in GE (× 500)
(from Galun & Lavee, 1966)

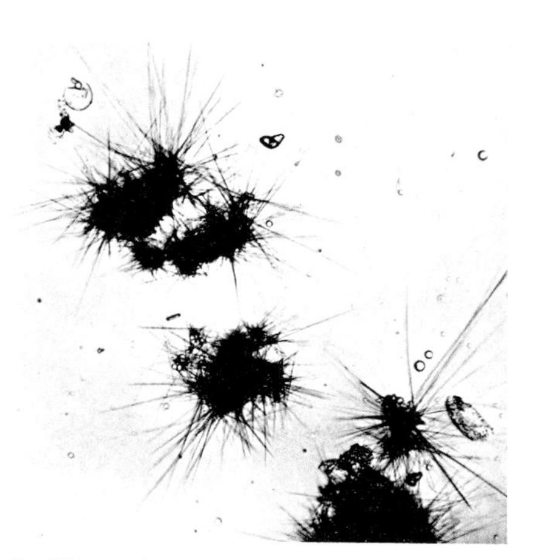

5. Olivatoric acid, in GE (× 125)

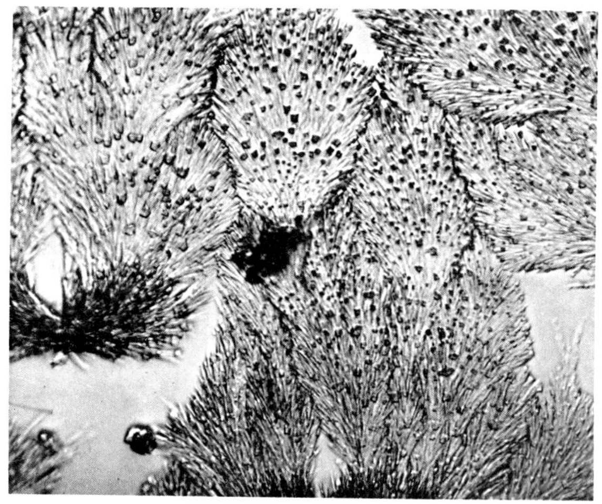

6. Evernic acid, in GE (× 500)

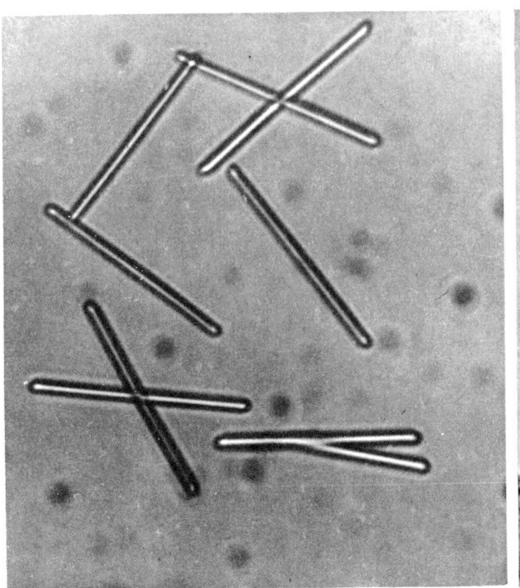

1. Atranorin, in GE (× 900) (from Galun & Lavee, 1966)

2. Lecanoric acid, in GE (× 1100) (from Galun, 1966)

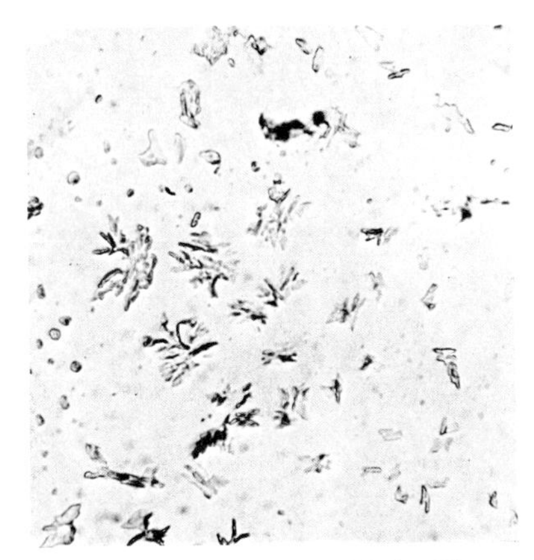

3. Fumarprotocetracic acid, in GE (× 450)

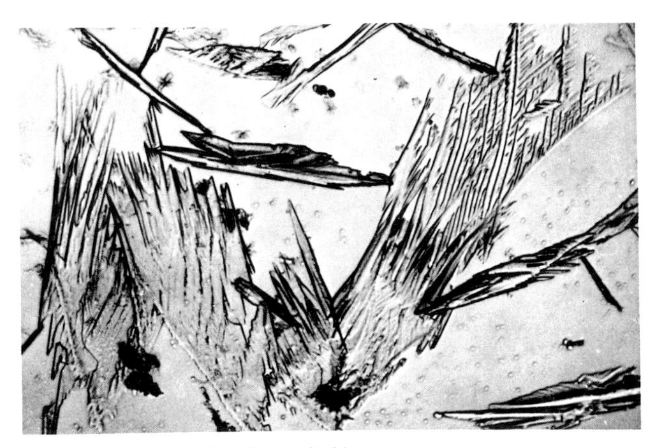

4. Usnic acid, in GAW (× 210) (from Galun, 1966)

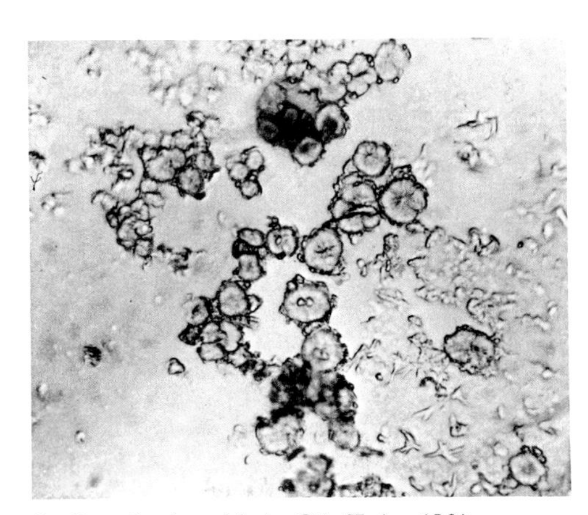

5. Gyrophoric acid, in GAoT (× 450)

6. Psoromic acid, in GAoT (× 400)

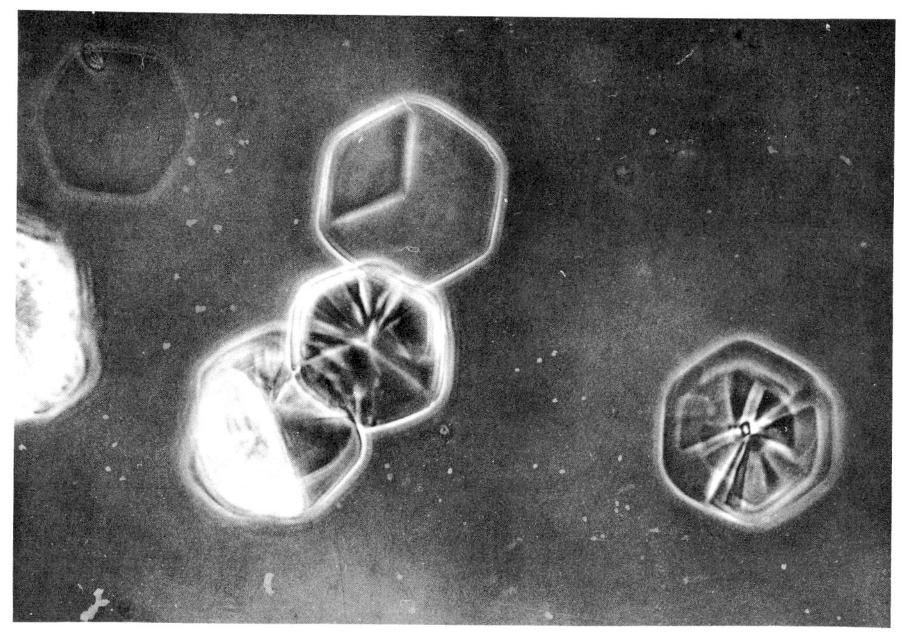

1. Stictic acid, in GAoT (× 480)

2. Salazinic acid, in GAoT (× 480)

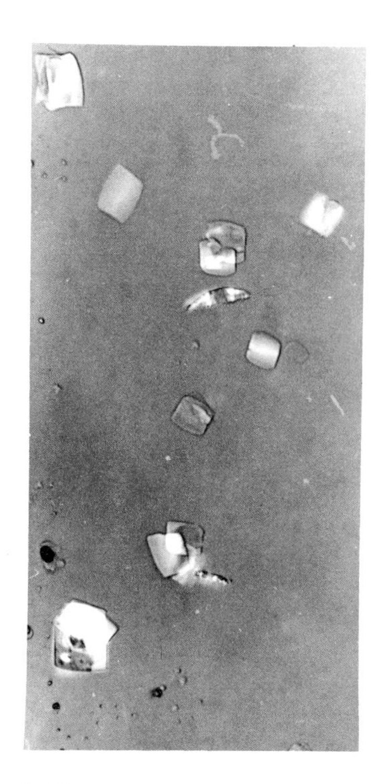

3. Norstictic acid, in GAoT (× 100)

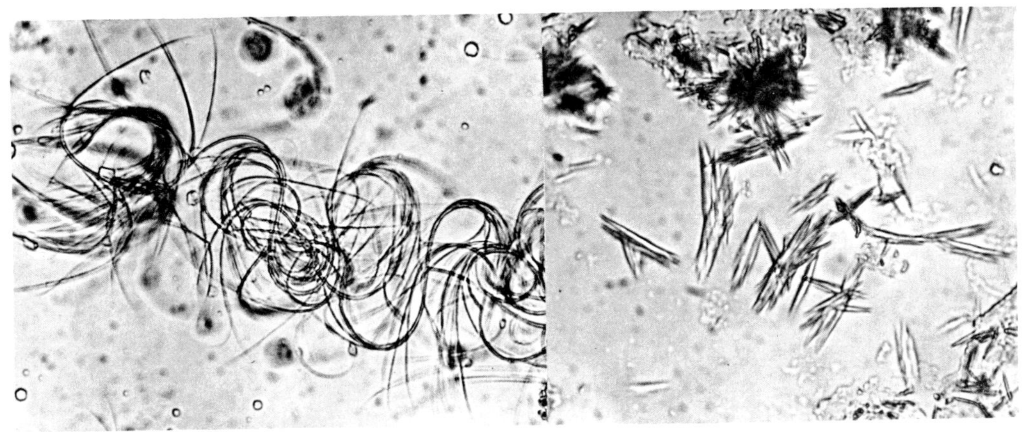

1. Chloroatranorin in GE (× 450) in GAW (× 450)

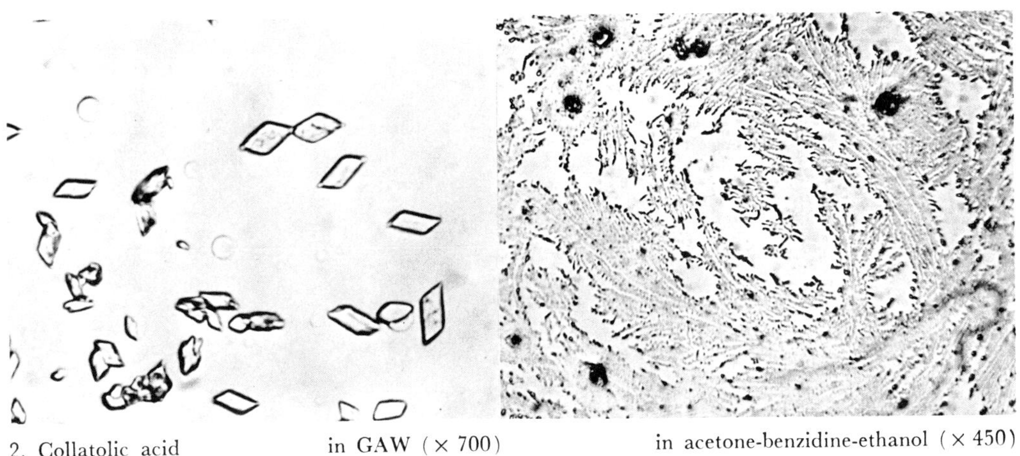

2. Collatolic acid in GAW (× 700) in acetone-benzidine-ethanol (× 450)

3. Nephromin, in GE (× 675)
(from Galun & Lavee, 1966)

PLATE XXVIII

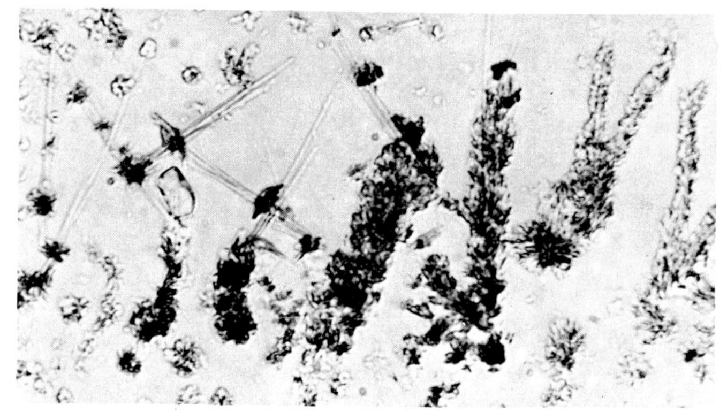

1. Stictaurin, in GE (× 400)

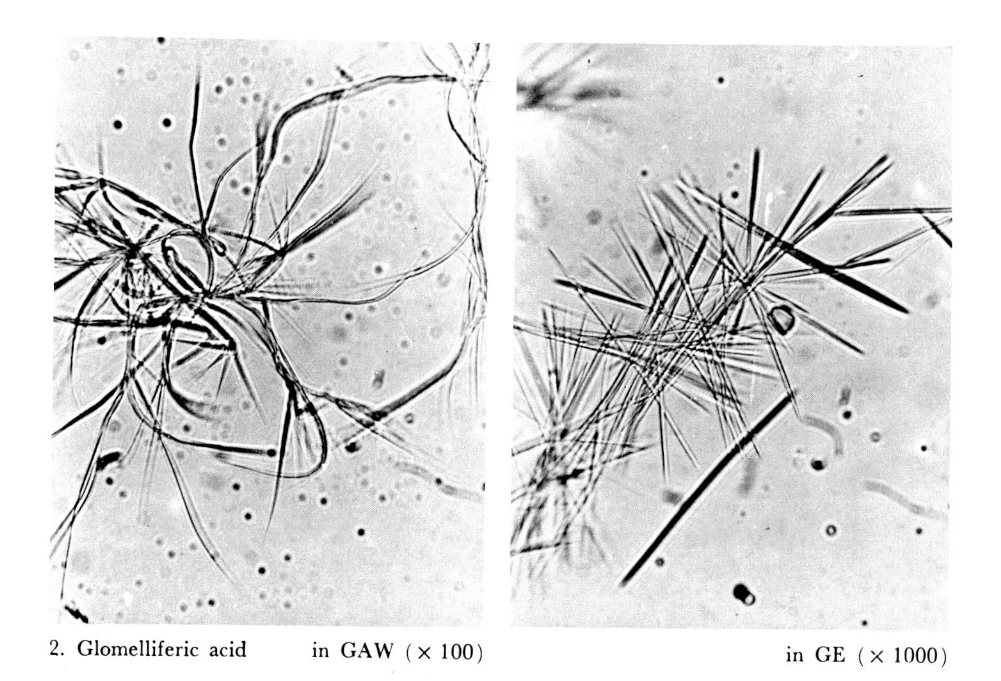

2. Glomelliferic acid in GAW (× 100) in GE (× 1000)

3. Parietin, in GE (× 125)
(from Galun, 1966)

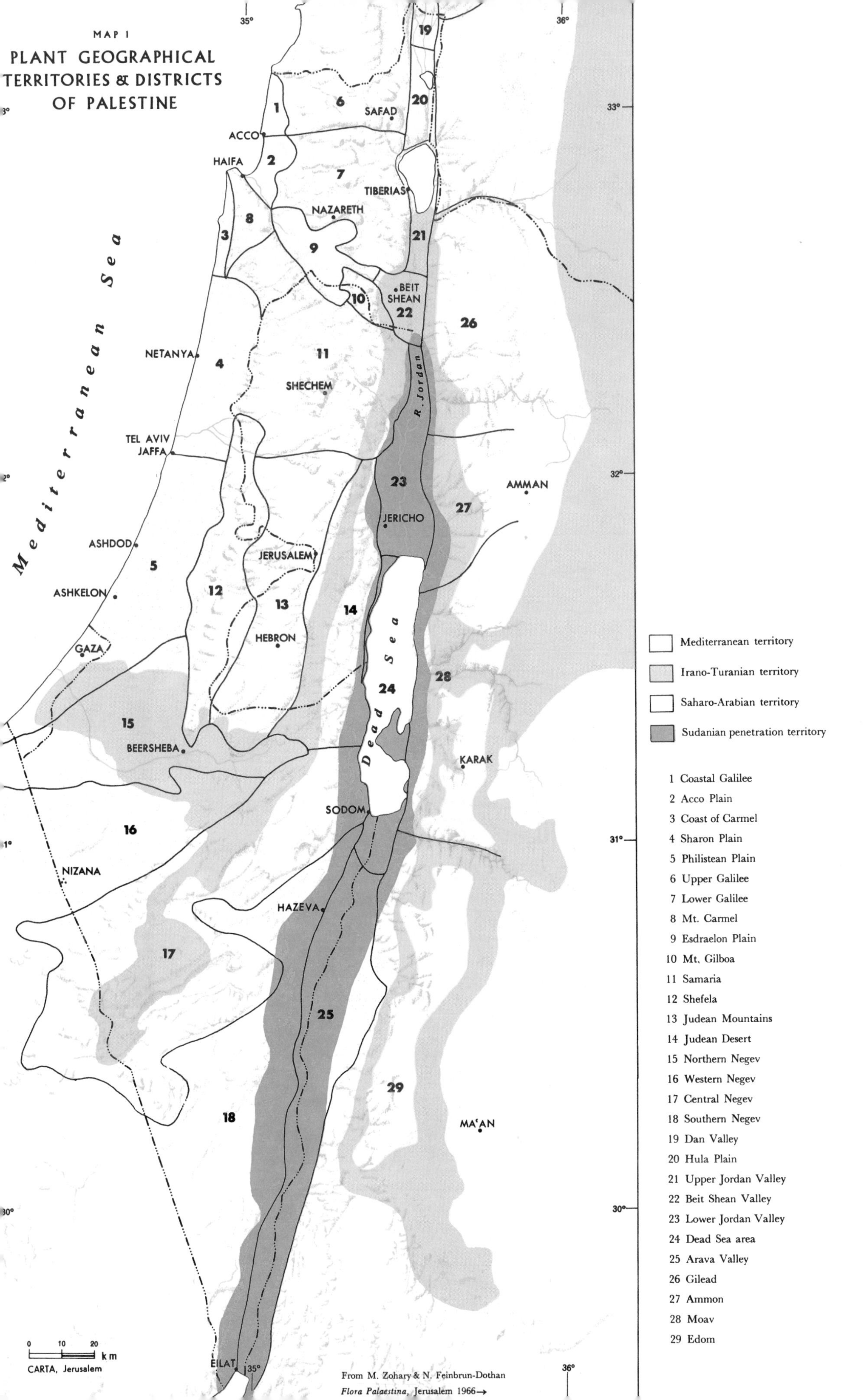

MAP I
PLANT GEOGRAPHICAL TERRITORIES & DISTRICTS OF PALESTINE
35°
36°
33°
32°
31°
30°
Mediterranean Sea
Dead Sea
R. Jordan
ACCO
HAIFA
SAFAD
TIBERIAS
NAZARETH
BEIT SHEAN
NETANYA
SHECHEM
TEL AVIV JAFFA
AMMAN
JERICHO
ASHDOD
JERUSALEM
ASHKELON
HEBRON
GAZA
BEERSHEBA
KARAK
SODOM
NIZANA
HAZEVA
MA'AN
EILAT
Mediterranean territory
Irano-Turanian territory
Saharo-Arabian territory
Sudanian penetration territory
1 Coastal Galilee
2 Acco Plain
3 Coast of Carmel
4 Sharon Plain
5 Philistean Plain
6 Upper Galilee
7 Lower Galilee
8 Mt. Carmel
9 Esdraelon Plain
10 Mt. Gilboa
11 Samaria
12 Shefela
13 Judean Mountains
14 Judean Desert
15 Northern Negev
16 Western Negev
17 Central Negev
18 Southern Negev
19 Dan Valley
20 Hula Plain
21 Upper Jordan Valley
22 Beit Shean Valley
23 Lower Jordan Valley
24 Dead Sea area
25 Arava Valley
26 Gilead
27 Ammon
28 Moav
29 Edom
0 10 20 km
CARTA, Jerusalem
From M. Zohary & N. Feinbrun-Dothan
Flora Palaestina, Jerusalem 1966→

MAP 2

PLANT GEOGRAPHICAL REGIONS REPRESE

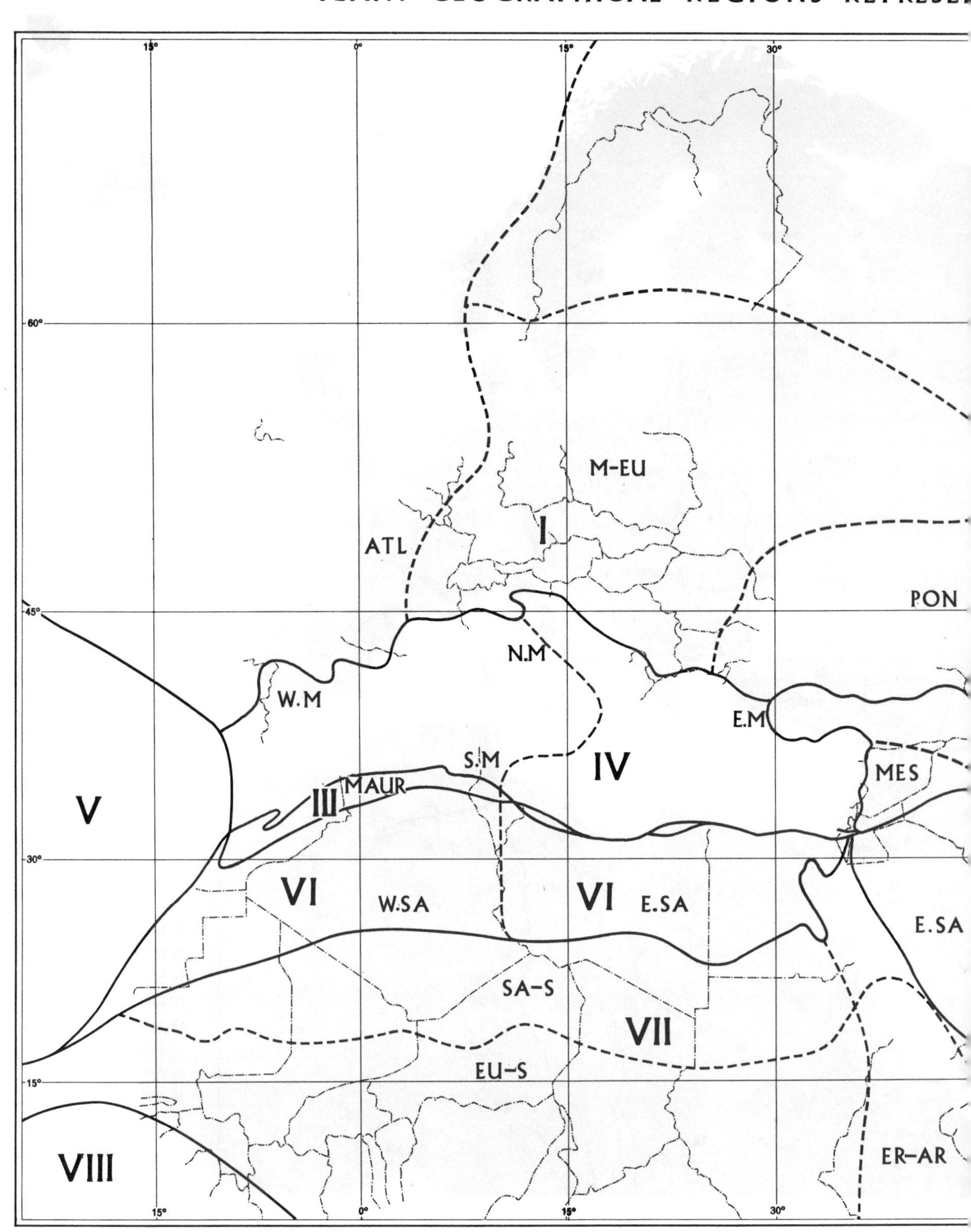

IN FLORA PALAESTINA

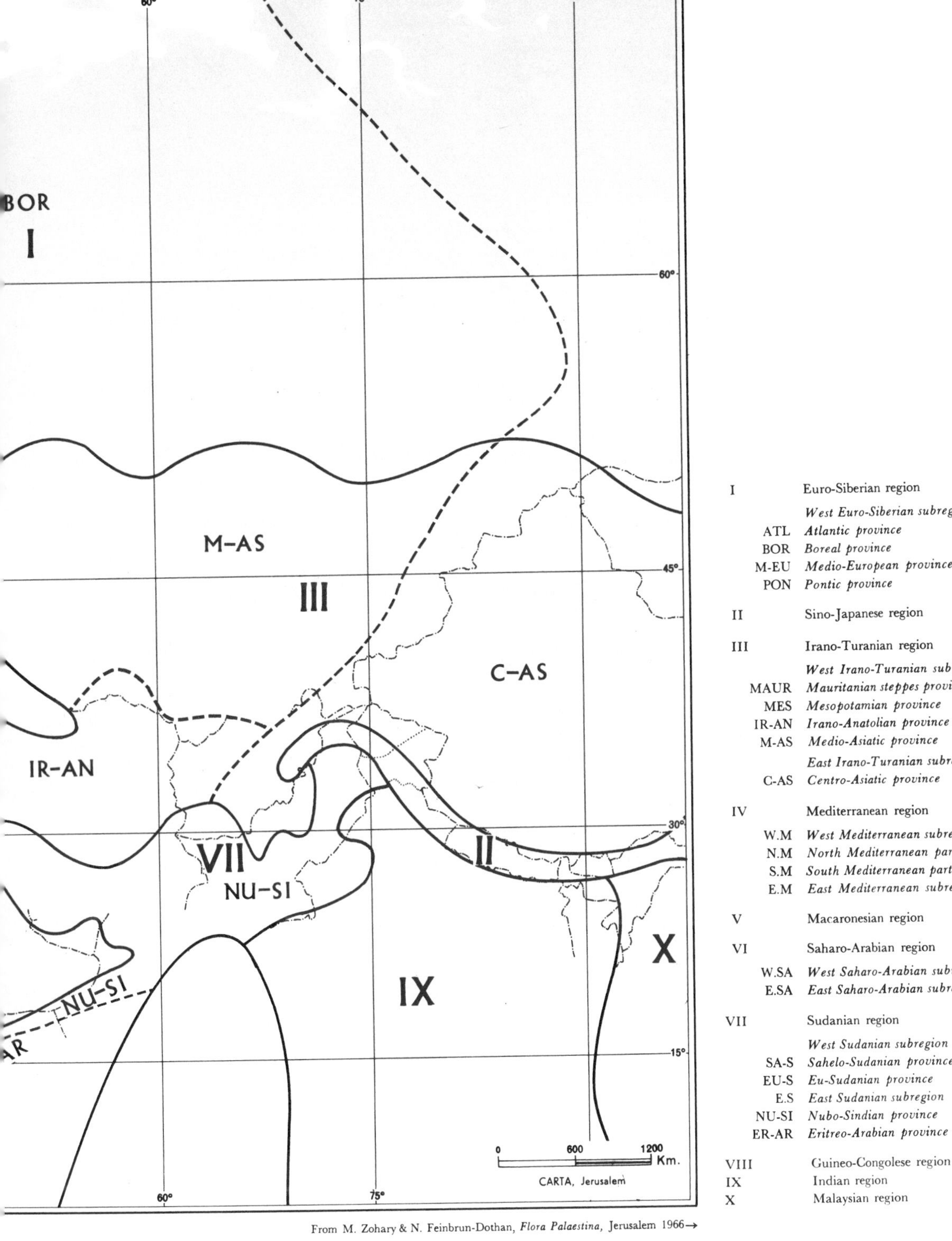

From M. Zohary & N. Feinbrun-Dothan, *Flora Palaestina*, Jerusalem 1966→

I — Euro-Siberian region
- *West Euro-Siberian subregion*
- ATL — *Atlantic province*
- BOR — *Boreal province*
- M-EU — *Medio-European province*
- PON — *Pontic province*

II — Sino-Japanese region

III — Irano-Turanian region
- *West Irano-Turanian subregion*
- MAUR — *Mauritanian steppes province*
- MES — *Mesopotamian province*
- IR-AN — *Irano-Anatolian province*
- M-AS — *Medio-Asiatic province*
- *East Irano-Turanian subregion*
- C-AS — *Centro-Asiatic province*

IV — Mediterranean region
- W.M — *West Mediterranean subregion*
- N.M — *North Mediterranean part*
- S.M — *South Mediterranean part*
- E.M — *East Mediterranean subregion*

V — Macaronesian region

VI — Saharo-Arabian region
- W.SA — *West Saharo-Arabian subregion*
- E.SA — *East Saharo-Arabian subregion*

VII — Sudanian region
- *West Sudanian subregion*
- SA-S — *Sahelo-Sudanian province*
- EU-S — *Eu-Sudanian province*
- E.S — *East Sudanian subregion*
- NU-SI — *Nubo-Sindian province*
- ER-AR — *Eritreo-Arabian province*

VIII — Guineo-Congolese region

IX — Indian region

X — Malaysian region